THE LAZY UNIVERSE

Cornelius Lanczos

From the archives of the Dublin Institute for Advanced Studies

THE LAZY UNIVERSE

An Introduction to the Principle of Least Action

by

Jennifer Coopersmith

Great Clarendon Street, Oxford, OX2 6DP,
United Kingdom

Oxford University Press is a department of the University of Oxford. It furthers the University's objective of excellence in research, scholarship, and education by publishing worldwide. Oxford is a registered trade mark of Oxford University Press in the UK and in certain other countries

First Edition published in 2017

Published in the United States of America by Oxford University Press
198 Madison Avenue, New York, NY 10016, United States of America

British Library Cataloguing in Publication Data
Data available

Library of Congress Control Number: 2016953473

ISBN 978–0–19–874304–0

Printed and bound by
CPI Group (UK) Ltd, Croydon, CR0 4YY

To
Cornelius Lanczos (1893–1974)
and
Murray John Peake (1952–)

"There's husbandry in heaven"

Act II, Scene I Macbeth

Shakespeare

Preface

The mathematician and physicist, Cornelius Lanczos (1893–1974), wrote one of the best books of physics explanation that has ever been written: *The Variational Principles of Mechanics.*[1] It explains the true meaning and philosophical content of the Principle of Least Action. The present book is a shorter and simplified version of that classic - but it is an original interpretation (anything taken directly from Lanczos is attributed and referenced by page number); it is also a paean to Lanczos.

Who is the book for? It is for *you*! It concerns a principle that underpins the whole of physics, so how could it not be important to understand even a tiny bit of it? The greatest theories have one common feature - they always bring in wisdom far-surpassing their original remit, extending into almost every walk of life. Therefore, one should not limit one's area of study but rather follow the maxim: "The more you know, the more you *can* know; the more you understand, the more you *can* understand."[2] Moreover; as well as an increased understanding of the world, you will be privy to a rare reward - to "theories of excessive beauty".[3]

It is true that the reader is assumed to have a background in the physical sciences, however a layreader could also read this book, with profit and enjoyment, by skim-reading the mathematics, or by reading just the Introduction (Chapter 1), the Final Words (Chapter 9), and the historical and popular chapters (2 and 8). Furthermore, the book is not a textbook: there are many equations, but the aim is to *explain* - why these equations and not others, and what do the equations mean?[4] The appendices are usually at a more advanced level and condensed in style, but they may be entirely passed over without loss of continuity in the main text. Apart from the invaluable asset of seeing how a problem is solved, the reason for so many appendices is twofold: to provide a

[1] Lanczos C, *The Variational Principles of Mechanics*, University of Toronto Press (1949). All page numbers will refer to the fourth edition, Dover Publications, Inc. New York (1970), and we shall write 'Lanczos, page x'.

[2] (author's maxim)

[3] Lanczos, page 229.

[4] This is in contrast to the approach in Synge and Griffith, Principles of Mechanics, 3rd edition, McGraw-Hill Book Company, Inc (1959), page 413, where the advice is: "Do not attempt to see a physical meaning in these [mathematical] operations; it will not help."

compact resource for the physicist (for example, a physicist on a beach holiday); and for the layreader to know what subject-headings to follow up at a later stage, if so desired. Enrichment material has been included but often as optional reading (in small font), or in parentheses, or footnotes. Also, we have avoided detailing all the qualifications, exemptions, and special cases, in order to be able to say things simply, and make bold statements.

Terminology: on the authority of both Lanczos, and Richard Feynman,[5] we refer to all the appropriate principles - Hamilton's Principle, Jacobi's Principle, Lagrange's Principle, and Maupertuis's Principle - as Principles of Least Action, as they are generically so. 'Variational Mechanics' means any physics problem that uses the Principle of Least Action. Also, the fact that we usually say 'least' as opposed to 'stationary' is explained in the main text - see Section 6.6.

The University of La Trobe is thanked for granting me an honorary research associateship, and for the use of library and computing facilities. It was Joe Petrolito, Emeritus Professor of Engineering at La Trobe, Bendigo campus, who introduced me to the website of Edwin F Taylor, Senior Research Scientist Emeritus at the Massachusetts Institute of Technology (MIT). This led to a short but rewarding email correspondence.

At OUP, I thank Sönke Adlung and Ania Wronski. The use of LaTeX, Ubuntu, Gimp, Wikipedia, Google, Metapost, Scilab, and Mfpic are gratefully acknowledged. Mal Haysom and Deborah Peake are thanked for their help with the diagrams (Deborah drew the maze, the encumbered porcupine and Stevin's 'Wreath of Spheres'). George Rogers, librarian at DIAS, once again anticipated my needs and supplied superb images of Lanczos and of Hamilton, with no fuss. Most of all, I am grateful to Gerald Jay Sussman, Panasonic Professor of Electrical Engineering at MIT, who undertook a critical reading of the book prior to printing. The errors he identified have been corrected. Finally, I thank Murray Peake for helpful discussions on maths and physics, and for leading me to Lanczos in the first place.

If I have inspired any reader to seek out the work of the master, Cornelius Lanczos, then it will be 'mission accomplished'.

[5] Feynman R P, *Feynman's Lectures on Physics*, Volume II, Chapter 19.

Contents

List of Figures

1

Introduction

It would be wonderful if there was one principle, simple to state, that could account for every process in the physical universe. But there *is* such a principle, a surprisingly well-kept secret, that accounts for *almost* every physical process. It is a principle that is more powerful than Newton's '$\mathbf{F} = m\mathbf{a}$', and a principle that doesn't have energy conservation as a requirement in every scenario. We know that Newtonian Mechanics must be replaced when speeds are very high, or the masses are tiny, or huge - but this 'new' principle still applies in these extreme regimes. How can one principle explain so much? The clue comes from the deep wisdom of the eighteenth-century French *philosophe*, Jean le Rond d'Alembert:[1]

> "*L'univers, pour qui saurait l'embrasser d'un seul point de vue,*
> *ne serait, s'il est permis de le dire,*
> *qu'un fait unique et une grande vérité.*"
>
> ("If one could grasp the whole Universe from one viewpoint,
> it would appear, if it is permitted to say this,
> as a unique fact and a great truth.")

No one knows whether d'Alembert's beautiful claim is correct but one thing is certain, if we cannot find one universal viewpoint then we will not arrive at one universal truth. For our viewpoint to be universal it is not a question of us all looking at the view from the same hilltop, rather, it is a requirement that all viewpoints are equivalent, and that there is just *one* universal rule or law or algorithm that solves the problem. Our new principle achieves this - but it seems incredible that one simple 'algorithm' could cope with all the specificity, variety, and complexity across the whole of physics. To make this plausible, we consider the following fable.

[1] D'Alembert, J le R, *Discours préliminaire de l'encyclopédie*, 1751.

The Lazy Universe. Jennifer Coopersmith, Oxford University Press (2017).
 DOI 10.1093/acprof:oso/9780198743040.001.0001

Figure 1.1 The suitors' puzzle.

There once was a King and he set a fiendish puzzle for prospective suitors who wanted to marry his daughter, the beautiful princess. The King had constructed a maze and the successful suitor had to provide the princess with instructions for collecting treasure from a casket. The young suitors had a few hours to look at a map of the maze and prepare their instructions. The King then looked at their answers and quickly whittled away the number of competitors to just two. These two suitors, Alfredo and Bruno, had very different approaches to the puzzle. Alfredo provided detailed instructions for every route, advising the princess to sight the tall column, visible from a distance over the hedges, to let her nose guide her to the fragrance of the frangipani tree, and also to listen out for the sound of bells chiming in the bell-tower, and water splashing at the fountain. Bruno came forwards with just a tiny scrap of paper on which it said, "Wherever you may find yourself, turn left at the next intersection. Eventually you will reach the casket." (The King had assured the contestants that there were no disconnected 'islands' within the maze.)

The King, a veritable sage, awarded the hand of his daughter in marriage to this second suitor.

The amazing thing about this suitor's instructions is that they are very simple to state and they are *universal* - they apply to any maze (although the maze must satisfy certain geometrical restrictions, for example, it cannot contain disconnected islands). There is also another curious attribute of Bruno's algorithm - it is *local*. The princess need only ever look as far ahead as the very next step (while always keeping an eye open for the casket). Although Alfredo's instructions may be broken down into steps the method is not local in the true sense (there are references to distant features - the sound of the fountain, the smell of the frangipani, the sight and sound of the bells in the bell-tower).

This is only a story but it demonstrates some essential points. In our new principle the method is truly local, and this is never the case for Newtonian Mechanics, even when a path may be broken down into lots of tiny incremental steps. But most astounding of all we shall find that the ensuing equations are *invariant*, taking exactly the same form, no matter what the scenario or what coordinates are used. It is not even necessary that the setting is time-independent, or that the components are passive (so the maze could writhe and undulate with time, and the princess could affect the maze, say, by trimming the hedges as she passed by). The reason for this invariance is that we have at last found something absolute: it is not a universal timepiece, yardstick, or reference frame, for there are none, it is a *principle*, and one that applies across almost every area of physics. We introduce it by way of a brief historical aside.

One of the most awe-inspiring developments in physics has been the shift from Newton's to Einstein's view of gravity. In Newton's Theory of Universal Gravitation, gravity is a force acting between bodies, however near or far. In Einstein's Theory of Gravitation (the Theory of General Relativity), the force is completely dispensed with. It is replaced by a patchwork of reference frames, sufficiently small, but seamlessly joined together, and, instead of responding to a force, the orbiting body now responds to *geometry* - the 'curvature' of 'space'. This is often represented heuristically by the image of, say, the Earth resting on a large two-dimensional surface, such as a trampoline, distorting this surface, and thereby affecting the trajectory of nearby small bodies, such as the Moon. These two theories - Newton's and Einstein's - are utterly different, and yet, amazingly, the experimental differences, for example, the predictions of the Moon's orbit, are practically nil. It turns out that Einstein's approach involves much more complicated calculations and, as we have just stated, barely any practical advantage - so why use it? The answer is that it has deeper explanatory power, it is applicable over a much greater range of problems, and it is philosophically more sound.

We have been talking just of gravitation, but the principle that we introduce explains not only gravitation but all kinds of problems in the physical sciences (statics and dynamics, optics, electricity and magnetism, quantum mechanics, physical chemistry, statistical mechanics, astronomy, materials science, hydrodynamics, quantum electrodynamics (QED), and so on); it also has deeper explanatory power, is applicable over a much greater range of problems, and is philosophically more sound. This principle is the Principle of Stationary Action (the PSA). It can be stated as:

> The Principle of Stationary Action
> *The physical system seeks out the 'flattest' region of 'space'.*

This is equivalent to choosing the 'straightest' possible path, which (usually) translates as the path of *least* 'distance'. One more thing, whether considering the 'flatness of space' or the 'straightness of paths', or the 'least distance', only the 'space' *nearby* - that is to say, locally, - needs to be inspected.

The subtitle of this book is the Principle of *Least* Action (PLA), but the principle just given is the Principle of *Stationary* Action (PSA)? 'Stationary' is a mathematical term meaning 'at a flat point of 'space' ' but whether that flat point implies a least path requires a further investigation - therefore the PSA is the more general principle, and incorporates the PLA. However, we shall find that the more stringent condition, the PLA, *is* the one we need, and later on we'll switch to calling our principle: the Principle of Least Action. We'll explain this in a later chapter (Section 6.6, Chapter 6).

Einstein's Theory of Special Relativity, that preceded his theory of General Relativity, starts with two postulates: 1) the laws of physics take the same form in every reference frame,[2] 2) the speed of light in a vacuum is a constant. These two postulates are strikingly different: the first postulate (the Principle of Relativity) is philosophical in character - Einstein coached us into realizing that physics just couldn't be practised unless postulate 1) applied. On the other hand, postulate 2) appears to be empirical - the speed of light is constant, yes, but perhaps, in another Universe, it might have been variable? Similarly, in the case

[2] When we say 'reference frame' we shall automatically mean a '*valid* reference frame'.

of the Principle of Stationary Action, the Principle has two postulates of very different natures. The first postulate, 1), we have already described - that the system 'space' is as 'flat' as possible, locally. This appears reasonable but rather abstract and philosophical; it sounds more like geometry than physics, and we need to know - 'flat' with respect to what? This is where the second postulate comes in, the one that contains the physical input. Postulate 2) states that what is actually being flattened is a certain specific physical quantity - '*action*'. This quantity has dimensions of *energy* × *time*, or *linear momentum* × *distance* or *angular momentum* × *angle*, and so on. In one of its first incarnations, 'action' was given as '$m \times v \times ds$', where m is the mass of a 'free' particle, v is its speed, and ds is a small distance along the particle's path. As we have to do with a postulate, we cannot justify the choice by deduction from even more elemental principles. Nevertheless, 'action' does seem like a worthy candidate for a telling physical quantity - it is a scalar (a pure magnitude, having no direction - therefore more likely to be an invariant), it 'spans the physical space' (nothing crucial is missed out), and it does so in the simplest way possible ($mvds$ is postulated rather than, say, $m^2v^3d^4s/dt^4$).

D'Alembert's "one viewpoint" implies objectivity, and this is difficult to arrive at in everyday life where prejudice abounds. For example: we barely notice the reaction-force against the soles of our feet, or on our bottoms, that is present almost every minute of our lives (Einstein teaches us that we are thereby not in 'free-fall', and so do not serve as a 'natural' frame of reference); on the other hand, in the rapidly rotating 'gravitron' at the funfair, we feel pinned as if by a great weight but have no sensation of our spinning motion (we merely notice that we can barely nod or move our arms). When revisiting a park that we knew as a child, we find that it resembles a pocket handkerchief rather than a vast estate - is the slight increase in the height of our eyes the source of this change? No, it arises because we have undergone an enormous (non-local) translation in time, during which our brain has totally altered. We watch the water sloshing about in a neighbour's swimming pool - perhaps they have installed a wave-generating machine? Upon closer inspection we find no machine, but realize that 'a giant hand' - an Earth tremor - is gently rocking the pool: therefore our initial assumption of an isolated system, defined by the edges of the pool, is wrong.

What helps us to achieve objectivity in physics (as opposed to everyday life) is the fact that we are bound by the strictures of mathematical tests. The PSA is centred on a mathematical test - a 'test of the flatness of 'space'' - which is remarkable for its ability to winnow away the distracting observer-dependent features and so arrive at the true invariant laws. It succeeds in this because it involves an 'extremal' feature of the mathematical landscape (something like 'shortest route between Peshawar and Kabul'), and these features are unique in that they don't depend upon the type of map or even the units used (the route is shortest whether we use a Mercator's or a Peter's projection, and whether we measure in feet or in metres). Before the test can be applied, we have to define what we mean by 'space'. We follow the historical development, and return to our discussion of Newton.

You most likely know Newton's Laws of Motion[3] but we want now to give a different perspective of them, emphasizing the philosophical assumptions. An implicit premise of Newton's Mechanics was the outstanding advance: 'space' and 'physics' are totally separate from each other. 'Physics' means forces, masses, and masses in motion. 'Space', following Descartes' invention of the coordinate system, means the three everyday space dimensions (commonly designated x, y, and z), and the time, t. All of x, y, z, and t are assumed independent of each other, and each goes on to infinity. Gone are the sixteenth century's tendencies, empathies, abhorrences, and vortices; and Newton's 'space' is empty, not full like Descartes'. (It is a void, which Newton does not abhor.)

Next after 'space' come particles - bodies with no internal structure but having an intrinsic property, mass. By Law (Newton's First), each 'free' particle is either at rest, or moves at constant speed and in a constant direction.

Finally, forces, **F**, are introduced, such as an attractive force between one particle and another. A force has one effect and one effect only - it causes a particle to accelerate. This is where mass plays its role, it determines how big the acceleration shall be, for a given force. (Apart from this, mass is inert - it doesn't depend on when or where the particle is, or on its state of motion.) All this is asserted in the Second Law, $\mathbf{F} = m\mathbf{a}$.[4] Another outstanding hypothesis was that for composite bodies (bodies made from many particles), or indeed for any complicated arrangement of particles, the net outcome could be obtained by 'summing' over the

[3] See Appendix A1.1, Newton's Laws of Motion.

[4] (But it could all have been so much more complicated; the force could have left the motion unchanged but caused the mass to swell, or it could have caused an acceleration not in line with **F**, or caused a third-order change in the position, and so on.)

influence of each particle considered on its own.[5] Thus was born the idea of the 'rigid body', an extended body made up from separate particles, but which could itself be treated as if it were one single particle, with all its mass concentrated at one point.

325 years on from Newton's "*Principia*"[6] it is hard to remain sufficiently impressed. As the philosopher, Schopenhauer, said, a theory passes from being rejected as ridiculous to being accepted but taken to be obvious.[7] Consider Newton's use of 'acceleration'. It was already known, from Galileo's Principle of Relativity, that uniform motion is relative to the observer. Newton turned this around: all non-uniform, that is to say, *accelerated*, motion is *not* relative to the observer, it *can* be known absolutely. Acceleration with respect to what? Answer, with respect to 'space'. But as the accelerations are absolute, then Newton's 'space' is absolute. So we have arrived back at Newton's wonderful abstraction, an infinite 'space', an inert and absolute background to physical happenings.

With the PSA, every tenet of Newton's Mechanics is challenged: the absolutes of Newton are avoided as every measure - whether it be a position, a speed, a direction, a time, and so on - is always defined with reference to something *within* the given system; action-at-a-distance does not occur - a global picture is built up, piece by piece, but by antennae which are sensitive only to conditions locally; the axes of 'space' no longer extend to infinity, are not necessarily independent of each other, and not necessarily independent of the masses within; and Newton's modular approach, building up complexity from more and more complicated arrangements of particles, each taken singly, is replaced by a holistic 'systems' approach.

Let's explain this 'systems' approach by analogy. Bertrand Russell (philosopher and mathematician) quipped that the activities of mankind amounted to the redistribution of matter within ±0.2% of the Earth, at its surface (this was before the era of space travel). To check Russell's claim, we could exhaustively track the motion of every single person, throughout recorded history, and note what masses they were carrying and where they deposited them; or, we could estimate the

[5] (Again, it could have been more complicated, the force might have been cast anew for, say, each trio of particles.)

[6] Newton, Isaac, *Philosophiae naturalis principia mathematica* (*The Mathematical Principles of Natural Philosophy*), 1687.

[7] (There is then the third phase, when the theory is again rejected.)

matter-content of cities built, crops grown, monuments constructed, bodies buried, and so on. In the second, 'systems', approach, we have lost the simplicity of basic elements (a person, their movements, what they are carrying) and instead have more abstract concepts relating to the whole system (cities, roads, pyramids, etc.). Some totally new possibilities arise ('the deceased') that were not catered for (!) in the first approach. We end up with a static tally of the mass distribution.

Another, more dynamic, example is given by the description of a football match. In the modular description we have only 'players' and 'a football'; in the other approach, the '*systems*-view', we have 'defence-position', 'attack', 'tackling', 'dribbling the ball', 'goal-kick', and so on. One counter-intuitive aspect of this systems-view is that we appear to have lost that quintessential feature of motion - its directionality. However, we soon realize that it is not lost but embedded in the whole-system structures (for example, 'goal-kick' has no absolute direction (say, 30° West) yet it conveys all the directional information required, and makes reference only to features within the system - the goal posts).

Returning now to the PSA, the method is as follows: (i) Instead of particles we have *individual components* of the system. These are chosen in a system-specific way. (They can be billiard balls, atoms, planets, lever arm, pendulum, capacitor, and so on, as the given problem demands); (ii) Instead of forces there are 'scalar structure functions', what in a previous life we have called the *energy functions* (the kinetic energy, and the potential energy); (iii) we identify all the independent 'motions' that the system can undergo. These 'motions' are the physical changes that happen naturally and that characterize the given system - what in a later life we shall call the 'degrees of freedom'. (For example, the planet orbits, the lever arm rotates, swings swing, and roundabouts turn.) (iv) We come finally to the application of a principle, a principle that requires an exploration of the 'space' in which the physical problem occurs. Knowing all the 'motions', we then choose an *alternative* set of 'motions' that *could* occur. These motions are hypothetical - we have hypothesized them - however we are not free to hypothesize anything we like: the 'motions' must all be in the same given 'space' (each system has its own 'system-space'), occur in the given time-window, and they must be 'nearby'. Now these 'motions' imply certain amounts of kinetic energy and potential energy consumed or generated in the given time. From these energies we compute a certain quantity - the total *action* - used up in the given time. In short, we determine the total hypothetical action for this choice of hypothetical motions. We then continue the exploration and consider another choice of hypothetical motions and again determine the consequential hypothetical action. And so on. The principle then asserts that of all choices for hypothetical motions, *the actual motions are those which make the change-in-action-between-choices come out to zero.* More evocatively, the system finely adjusts itself, via the actual motions (acting in concert but instant by instant taking their marching orders from the scalar structure functions) in just such a way that the action used is least. That these subtly different versions - the italicized one and the evocative one - are the same will emerge during the course of this book.

Be reassured: these ideas are new and many and abstract; there is no way they can be understood in one go. It is useful to collect them together in one place, but not possible to convey all the nuances in a single paragraph. For example, we shall later on discover that sometimes the 'space' exploration is made explicitly by us (in the method known as the Principle of Virtual Work) and sometimes the mathematics takes care of it (in the Variational or Lagrangian Mechanics). Also, there is sometimes an elision made between the 'motions' as 'degrees of freedom' and the 'motions' as hypothetical 'variations'.

Did we write 'hypothetical'? Yes, this is the *piece de resistance*: the 'system-space' is a *virtual* abstract space, and this is what finally enables us to achieve the required objectivity (the actual physical space could have this, that, or the other observer-bias, whereas the virtual abstract space is neutral).

Here is a summary of the main virtues of the PSA.

(i) It does the job.

(ii) As forces play no part in the method then 'forces-of-constraint' also play no part. Incredible but true.

(iii) Better understanding of physics. We can now have 'cat' and 'mouse' instead of only 'particles' and 'particle-particle interactions'. But a 'cat' is more than the sum of its 'particles'.

(iv) No hard and fast distinction between 'active' and 'passive' components. (Just as a river carves out the river-bed, and the river-bed determines the path of the river, so the 'curvature of space' affects the paths of bodies, and moving bodies affect the 'curvature of space'.)

(v) Philosophically superior: local (and this is all we ever detect experimentally); no pre-existing empty 'space' (just take the world as it is and then examine it[8]); the system is what's important.

(vi) Global View. Even when 'space' is flat locally, 'curvature' can still arise globally - by patching together smaller regions with the requirement that there is no 'puckering at the seams'.

(vii) The PSA gives prominence to energy and to the whole system. Kinetic energy is shown to be a more fundamental and primitive concept than force. Also the dichotomy between kinetic energy and potential energy is explained.

(viii) The PSA reveals a deep connection between symmetry and conserved properties. (Newton's Mechanics does not lead naturally to any conservation laws except for one, the conservation of linear momentum.)

(ix) In Newtonian Mechanics no attempt is given to show how robust the solution is (how it changes following small changes in the starting conditions) or to give ball-park estimates. The PSA, via Hamilton's Mechanics, does address these questions.

(x) Amazing unity of approach across almost the whole of physics. (Almost? The exceptions will be discussed in due course.)

Apart from the fact that the PSA is the keystone of physics, and therefore an indispensable tool for the professional engineer or physical scientist, there are two other reasons why we would like to awaken an appreciation of it, even a non-mathematical appreciation. The first is a pragmatic reason. We all know, roughly speaking, what space, time, mechanics, quantum mechanics, matter, and energy are about. These ideas have passed into the public domain. It would be inefficient to start from scratch in our science classes, and not even incorporate advances made during the seventeenth and eighteenth centuries. Somehow the PSA has got missed out - it is time to correct this. The second reason is aesthetic. The beauty of physics does not reside only in the beauty of the night sky, a rainbow, or a sunset. It resides even more in the interior logic, the principles which reach across vast areas of the physical

[8] This is more correct, especially when we remember that all our observations really have been carried out in the presence of large gravitating masses. Even where experiments are carried out in remote regions, the results still need to be brought back to Earth - in our present state of evolution.

world, unifying them into a self-consistent whole, and with the most economical set of starting premises. (One could call it the 'Aha' feeling.)

Alice (from Lewis Carroll's *Alice's Adventures in Wonderland*) tried so hard to get through the door in order to see the exquisitely beautiful garden beyond. Consider this garden as a metaphor for physics: it's true that without mathematics we shall never be able to wander freely around this garden, but it would be wonderful indeed if we could just be lifted up to peer at it through the keyhole.

2

Antecedents

A new principle rarely arises completely out of the blue, there are usually vague presentiments in the air. Some precursors to the Principle of Least Action are described (we don't attempt to give an exhaustive history).

Simon Stevin (1548–1620)

Stevin (Stevinus) was a 'geometer' from Bruges in Flanders. The discovery of which he was most proud concerns the condition for static equilibrium of weights on inclined planes (see Figure 2.1):

What relation must hold between the masses and the lengths of the triangle in order for the system to be in equilibrium? Our first guess might be to 'resolve the forces' into components, then balance the horizontal components to zero, and, finally, equate the vertical components to the weights. Stevin, in Antwerp in 1588, attacked the problem in a different way - (he was working 200 years before vectors, and 99 years before Newton's forces).

The first remarkable thing Stevin did was to bring (hypothetical) motion into this static set-up. He recast the problem and imagined a chain of spherical masses that could circulate around the triangle, with the lower part of the chain hanging freely. Stevin idealized this arrangement: the chain could move in either direction, without friction, and would never get stuck on the pointy bits (imagine a little frictionless pulley at the apex of the triangle). Also, the mass had to be *uniformly* distributed along the chain - think of a bead chain like the ones used today to open and close Venetian blinds. Stevin argued that if the section of chain from A to X pulls more than the section from X to B then the chain will over-balance to the left. It will circulate anticlockwise (excuse the anachronism) until the original portion AX then occupies the arc Y to A. The entire length of chain that hangs freely below the little table, between A and B, can be ignored - by symmetry it is always in balance

The Lazy Universe. Jennifer Coopersmith, Oxford University Press (2017).
 DOI 10.1093/acprof:oso/9780198743040.001.0001

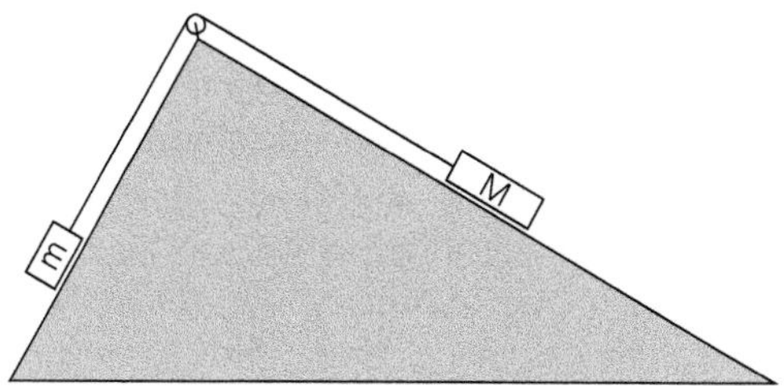

Figure 2.1 Weights, connected by a cord, on inclined planes.

Figure 2.2 'Wreath of spheres' draped over inclined planes (after Stevin).

with itself. However, the replacement section of chain now lying along AX will again over-balance that new section now lying along XB. In other words, the new state, after circulation, is identical to the initial state. But this will always be true (the new state, after circulation, will always be identical to the initial state), and so the chain will always over-balance, and will keep circulating for ever.

But a continuously circulating chain, without an engine to drive it, is absurd. This is the second remarkable step taken by Stevin - he considered 'perpetual motion' as self-evidently absurd, and used this as the basis for an argument (a *reductio ad absurdum* argument).[1] To avoid the absurd outcome then the starting premise - section AX over-balances section XB - must be wrong; in other words, the initial distribution of masses, with the chain draping itself smoothly over the surfaces, must already be in equilibrium. This is a powerful proof, as it applies for an infinity of different starting states (different inclined planes).

Stevin realized he had discovered an eternal truth, and he displayed a diagram of his 'Wreath of Spheres' (*clootcrans*) as the frontispiece of his book on mechanics. The punchline, from our point of view, is that: without the use of forces, considering only gentle movements of the chain in harmony with the constraints (motion within certain surfaces), Stevin showed that, with respect to these motions, nothing changes, and equilibrium is maintained. In fact (going beyond Stevin's own interpretation), the proof can be made to sound like pure geometry: for equilibrium, it is only necessary that the length of chain along an inclined plane is equal to the length of that inclined plane. Moreover, as any bumpy surface can be thought of as lots of tiny triangles, we have only to lie a uniform chain on this surface, with no bunching up or stretching out, and it *will* be in equilibrium.

Christiaan Huygens (1629–95)

Huygens, in Paris in 1656, tried to understand billiard-ball-type collisions but without the use of forces (this was some thirty years before Newton's *Principia*, but billiards was a game that had been played since the fifteenth century). Then, as now, scientists were motivated by the thrill of showing that a great name was wrong. Descartes was the towering authority in the middle of the seventeenth century, and we shall see that three scientists (natural philosophers) in our story found errors in Descartes' work - Fermat, Huygens, and Leibniz.

Back to Huygens. In his rules of collision, Descartes had claimed that a small body, hitting a larger body at rest, would never be able to shift it. Huygens knew that this couldn't possibly be right - it didn't agree with

[1] This is probably the first time this had ever been done - that is, the use of the impossibility of perpetual motion as the basis of an argument.

experiment.[2] To show that Descartes was wrong, he used a remarkable demonstration. (Huygens might have been a bit apprehensive about contradicting the Cartesian view - Descartes had been a regular visitor to the Amsterdam home of Huygens's parents, and his tutor was horrified to hear of the young Huygens' 'heresy'.)

Huygens's remarkable demonstration consisted in putting the Principle of the Relativity of Motion (due to Galileo) to *quantitative* use - the first time this had ever been done.[3] He imagined a 'billiard-ball collision' viewed simultaneously from two vantage points - from a smoothly coasting canal-boat, and from the canal-bank (undoubtedly his childhood in Amsterdam was an influence). Here is a picture of this thought-experiment, taken from the frontispiece of Huygens's treatise.

It's not clear from this picture whether the experiment was carried out on the boat or on the canal-bank. It doesn't matter, so let's say it happened on the boat. On the boat, the large mass hits a stationary smaller mass. Now we are free to choose the speed of the canal-boat

Figure 2.3 Huygens's canal-boat thought experiment[4] (from *De Motu corpore ex percussione*, 1656)[5].

[2] Descartes also knew that experiment was contradicted, but he had the perfect fudge - see Coopersmith, J, *Energy, the Subtle Concept*, Revised Edition, Oxford University Press, (2015) - hereafter referred to as Coopersmith, *EtSC*.

[3] (to the author's knowledge)

[4] The figure shows balls of the same size, yet we can imagine that one is more massive than the other.

[5] Huygens C, *De Motu corpore ex percussione*, 1656, published by Martin Nijhoff NV.

so that it will exactly cancel out the incoming speed of the large mass when viewed from the bank. Then, as viewed from this bank, the small mass will collide with a stationary larger mass, and then *cause this larger mass to move* - and therefore Descartes is proved wrong.

The Relativity of Motion has been used to veto certain outcomes, but Huygens recognized that a general principle was at work: only those rules are correct which guarantee the same outcomes, no matter what frame of reference they are viewed from. The punchline: as with Stevin, certain 'motions' (in Huygens's case, uniform motion of one reference frame with respect to another) have resulted in 'no change'.

Huygens went on to use other symmetry arguments to develop his own rules of collision,[6] and these were in better accord with experiment than Descartes's rules. Although symmetry arguments alone didn't answer to all possible outcomes, the telling point, for us, is that so much could be explained this way, and without the use of forces. Huygens noted in passing (drawing upon both his collision theory, and his theory of the compound pendulum) that a certain quantity was conserved: it was the total mv^2 ... (m was the mass, and v the speed, of each body).

Gottfried Wilhelm Leibniz (1646–1716)

Huygens attached no great significance to the quantity mv^2. Leibniz, on the other hand, immediately realized its importance. He already knew of it from Galileo's work on free-fall,[7] but when he happened to learn of its conservation from Huygens (Huygens and Leibniz met regularly at the *Academie Royale des Sciences* in Paris) Leibniz built a whole new philosophy around it, calling it *vis viva* or 'live force'. (Leibniz, from Hanover, introduced other 'whole new philosophies', fully justifying the German epithet, *universalgeni*.) Straightaway, Leibniz made big play of how his new 'live force', mv^2, trumped Descartes's 'quantity of motion', mv. (Leibniz's paper was called "A brief demonstration of a famous error of Descartes and other learned men, concerning the claimed natural law according to which God always preserves the same quantity of motion; a law which they use incorrectly, even in mechanics";[8] the title, at any rate, was anything but brief.)

[6] Coopersmith, *EtSC*, Chapter 3.

[7] Galileo found that v^2 (and therefore mv^2) was proportional to the fall-height.

[8] Leibniz G W, "Brief demonstration..." (1686) in Philosophical Papers and Letters, ed Loemker, Chicago University Press (1956).

Leibniz's mv^2 was essentially the same as our modern kinetic energy, $\frac{1}{2}mv^2$ - it had no dependence on direction, it obviated the need to calculate accelerations, and, above all, it encouraged a 'systems' view. It was slowly learned, over the next 200 years, that energy existed in various scalar forms (the 'structure functions' of Chapter 1), and that altogether, within a closed *system*, it was conserved.[9] We add that Leibniz's philosophy also stressed another new idea (in common with Huygens) - the importance of symmetry: a free particle could not swerve to the left or the right as it had no 'reason' to do so (Leibniz's Principle of Sufficient Reason). Leibniz was hundreds of years ahead of his time in using such arguments.

The dispute between Leibniz and the Cartesians evolved into a controversy between the Leibnizian school and the Newtonian school, between a whole-system view and an individual-particle view, and between 'kinetic energy' and 'force'. Ultimately (as we shall find), it evolved into a contest between Newtonian Mechanics and the Principle of Least Action, a contest which, some say, still persists today.

Maximal/minimal properties

At the same time as conservation principles were beginning to hold sway, an alternative approach was coming in: nature was economical as well as conservatory with her resources. What was the measure of nature's resources - time of travel? length of path? the total potential energy? Even while this was still unclear, it seemed to some philosophers self-evident that optimization was a crucial driver of physical processes. Chief among these natural philosophers was the great Swiss mathematician, Leonhard Euler (1707–83), who wrote:

> "nothing happens which has not some maximal or minimal property."[10]

We shall meet him again in this chapter.

Max/min problems were already known about in antiquity. There is the story about the Phoenician princess, Dido. After running away from home, she reached the coast of North Africa and tried to buy some land.

[9] Coopersmith, *EtSC*.

[10] Euler L, '*Additamentum I de curvis elasticis*', 1744, English translation in Oldfather et al, Leonhard Euler's elastic curves, Isis 20 (1933) pp 72–160.

Legend has it that she was allowed to purchase only as much land as could be encompassed by the hide of a bull. From this unpromising start, Dido cleverly maximized her land-area: she cut the hide into very thin strips and joined these end to end; she chose a plot by the sea so that the coastline would be part of the boundary; and she shaped the long strip of hide into the optimum shape - a semi-circle, up against the coast.

Heron of Alexandria (around AD 60) found that light reflected from a plane or curved surface took the shortest distance between the light-source and the detector (not counting missing out the mirror altogether). And Pappus of Alexandria, around 300 AD, observed that beehives had that special shape, hexagonal, which could hold the greatest volume of honey for the smallest expenditure of wax.

Our story becomes quantitative in the mid-seventeenth century with Pierre de Fermat (1601–65), French lawyer and 'amateur' natural philosopher. Fermat agreed with Heron's results for the reflection of light but when it came to refraction he found that the path was not one of shortest distance but of least time. He elevated this to a general principle, his Principle of Least Time:[11]

> "Nature operates by means and ways that are easiest and fastest."

Fermat was delighted with his Principle but "astonished"[12] because his laws of refraction, while completely ageeing with Descartes's laws, started from utterly opposite premises: Descartes (Fermat) required that light travelled faster (slower) in denser media.[13] The Cartesians were not impressed - "The shortness of the time? Never."[14] By the way, we mentioned symmetry arguments in connection with Huygens, and Leibniz, but they are also present in Fermat's Principle: for the reflection of light, the angle of incidence is *equal* to the angle of reflection.[15] Also, another feature in Fermat's Principle (the importance of this will be recognized much later), is that the correct path is not only the one taking

[11] (in 1662, in a letter to de la Chambre) Goldstine H, *History of the Calculus of Variations from the Seventeenth through the Nineteenth Century*, Springer-Verlag, New York (1980).

[12] Dugas R, *A History of Mechanics*, Dover Publications Inc. (1988).

[13] Imagine how difficult it would have been to measure these speeds - or even know that light has a speed.

[14] Dugas R, as above.

[15] A symmetry rule applies in refraction as well: $n_i \sin\theta_i = n_r \sin\theta_r$.

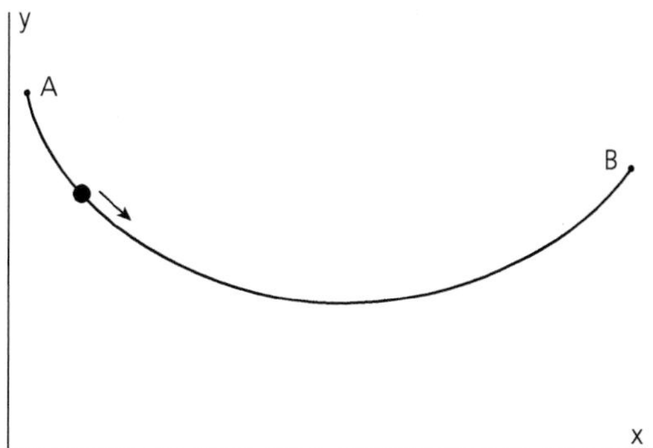

Figure 2.4 *Brachystochrone*, curve of 'swiftest descent' (schematic).

the least time, but the one whose time-of-travel is *the same* (to first-order) as the time-of-travel for neighbouring paths.

Isaac Newton (1643–1727), in his landmark work, familiarly known as "*The Principia*",[16] derived the optimum shape for a solid body so that it would move with the least resistance through a fluid. However, apart from this, and the '*brachystochrone* question' - see below, there are no other examples where Newton tackled such max/min problems.

The Bernoullis of Basle in Switzerland exhibited an extremal property all of their own - they had the largest number of mathematical geniuses in one family (eight) that has ever been recorded. The brothers, Jakob and Johann Bernoulli, were at the start of it (born in 1654 and 1667 respectively), both brilliant but very competitive - they baited each other with silver ducats and impossible deadlines for the solution of difficult problems.

One famous problem was the search for the *brachystochrone*, the path of 'swiftest descent', taken up by Johann Bernoulli in 1696. A weight slides down a curve, and the question is: what shape must the curve have so that the travel time between given start- and end-positions is least? (Assume no friction, and that the destination point is not dead vertically down from the starting point.) Galileo Galilei (1564–1642) had already noted that the quickest path was *not* along the straight line connecting the points.

[16] Newton, Isaac, *The Mathematical Principles of Natural Philosophy*, (1687) translated 1729 by Andrew Motte.

Leibniz, Newton, de l'Hôpital, Tschirnhaus, and both Bernoullis all came up with solutions for this swiftest-descent curve. Newton's had been submitted anonymously but Johann "recognized the lion by its claw".[17] Jakob and Johann both solved the problem but employed very different methods, each opening up grand new vistas in physics.

Jakob's method was complicated and messy, but showed the way, ultimately, to more general techniques for max/min problems, a forerunner of Euler's and Lagrange's 'calculus of variations' (see Section 3.7, Chapter 3).

Johann's method was a masterpiece of lateral thinking: he considered that the falling mass passed through successive sheets of 'denser and denser gravity', and was 'refracted' at each boundary in such a way that its total journey-time was least (a smooth curve was obtained by imagining thinner and thinner sheets). This was analogous to Fermat's least-time path for light. Thus, Johann's insight foreshadowed both Hamilton's conjoining of light and matter into one theory (Chapter 7), and Einstein's theory of gravitation (in which a freely-falling mass moves along a curved path, as dictated by the curvature of 'space' - the 'gravitational refractive index'). Johann was centuries ahead of his time.

What was astounding for the contemporary mathematicians - and ever since - was that the *brachystochrone* turned out to have exactly the same shape as Huygens's *tautochrone*,[18] and both curves were the same as a third curve, the cycloid.[19] As Johann Bernoulli wrote,

> "...you will be struck with astonishment when I say that this very same cycloid, the *tautochrone* of Huygens, is the *brachystochrone* we are seeking."[20]

Before we move on from these wonderful old 'chrones, the punchline is: whether 'least time' or 'same time', the optimized path is *stationary* - the total travel-time doesn't change with respect to *small* variations in

[17] Kline M, *Mathematical Thought from Ancient to Modern Times*, Oxford University Press (1972).

[18] The *tautochrone* is the curve guaranteeing equal travel-times irrespective of (modest) variations in the starting position.

[19] A cycloid is the curve traced out by, say, a pebble stuck in a tyre as a bicycle rolls forwards.

[20] Kline, as above, page 575.

either the whole-curve-shape or the start-time. (The term 'stationary' will be explained in Chapter 3.)

There were other curves that could be determined by the max/min method, for example, the curve of a hanging chain or suspension bridge, known as the catenary; and the '*elastica*', the shape of a flexed metallic band (think of a metal ruler). This latter problem was investigated by Daniel Bernoulli (1700–82), the middle son of Johann, and the first true physicist (as opposed to mathematical physicist).[21] The *elastica* could be determined, but what actually was the property being maximized or minimized? The following correspondence between Daniel Bernoulli and Leonhard Euler, from 250 years ago, makes fascinating reading:[22]

Bernoulli to Euler, 5 May 1739:

"I have today a quantity of thoughts on elastic [metal] bands... I think that an elastic band which takes on of itself a certain curvature will bend in such a way that the live force will be a minimum, since otherwise the band would move of itself. I plan to develop this idea further in a paper; but meanwhile I should like to know your opinion on this hypothesis."

Euler's reply, 5 May 1739:

That the elastic curve must have some maximum or minimum property I do not doubt... but what sort of expression should be a maximum was obscure to me at first; but now I see well that this must be the quantity of potential forces which lie in the bendings; but how this quantity is to be determined I am eager to learn from the piece which your Worship has promised."

Bernoulli replies, 1742:

My thoughts on the shapes of elastic bands, which I wrote on paper only higgledy piggledy and long ago at that, I have not yet set in order."

and again, from Bernoulli to Euler, 1743:

"May your Worship reflect a little whether one could not deduce the curvature... directly from the principles of mechanics... I express the potential live force of the band by $\int ds/r^2$... Since no one has perfected

[21] Coopersmith, *EtSC*, Chapter 7.

[22] Truesdell, C, introduction to *Leonhardi Euleri, Opera Omnia*, 2nd Series, Vols X and XI, Fussli, pp 173–4.

the isoperimetric method[23] as much as you [Euler], you will easily solve this problem of rendering $\int ds/r^2$ a minimum."

And in his famous paper the following year (1744) Euler writes:

> "although the curved shape assumed by an elastic band has long been known, nevertheless the investigation of that curve by the method of maxima and minima [could not be carried out until the] most perspicacious Daniel Bernoulli pointed out to me that the entire force stored in the curved elastic band may be expressed by a certain formula, which he calls the potential force, and that this expression must be a minimum in the elastic curve."[24]

(In other words, Daniel needed help with the maths and Euler with the physics.) What is so fascinating is that we are witnessing not only the birth of variational mechanics, but also of kinetic energy (live force) and potential energy (potential live force). However, all these max/min problems were tackled individually, and it was not clear what if any overriding principle might encompass them all.

The Principle of Virtual Work

This principle will have a chapter to itself (Chapter 4). The first glimmerings of it, in the guise of the Principle of Virtual Velocities, are found in Aristotle's analysis of the lever. When the lever is in balance, then the end with the heavier (lighter) load moves slower (faster). But, one may ask, if the lever is balanced, then the ends don't move at all? Yes, that's why these movements are called *virtual*, or 'mathematically imagined'. This mathematical experiment is carried out in order that the criterion for balance may be established. (This will be explained in Chapters 3 and 4.) The same Principle was taken up by Stevin: "What is gained in the force is lost in the velocity", and then again by Galileo. Galileo made an important advance in recognizing that it is only the velocity *in the direction of the force* which counts.

Finally, Johann Bernoulli, ever the clairvoyant, took up the Principle and realized its potential to solve all the problems of static equilibrium. (Before Johann, the Principle was limited to cases where there

[23] A perimeter of given fixed length enclosed an area: what was the shape of the perimeter-curve needed to maximize the area?

[24] Euler L, letter of September 1738 to Daniel Bernoulli, in *Die Werke von Daniel Bernoulli*, Band 3, *Mechanik*, ed Speiser, Birkhauser Verlag, (1987) p 72.

were just two forces, the 'moving force' and the 'load'.) Bernoulli made significant improvements and took the Principle to a new level of generality:

(i) he applied the Principle to any number of forces - as many as the system demanded - (all in static equilibrium).

(ii) he no longer assumed that the virtual velocity was in inverse proportion to the force but, rather, considered the product of force and 'virtual velocity in the direction of that force'. He called this product the '*energy*'.

(iii) he adopted a sign convention as follows - if the angle between force and velocity was acute (obtuse) then the product had a positive (negative) sign. (In modern parlance, the sign convention was the usual one for the 'scalar product' of two vectors.)

(iv) the criterion for equilibrium was that all these products, with appropriate signs, had to sum to zero.

Bernoulli's revolutionary idea, a watershed in physics, was never published, but was written down as some throwaway remarks in a letter to Varignon in 1715. Fortunately, Varignon didn't throw them away:[25]

> "In every case of equilibrium of forces, in whatever way they are applied and in whatever directions they act on [one] another, mediately or immediately, the sum of the positive energies will be equal to the sum of the negative energies taken positively."

This is the first time that the word 'energy' appears in physics (an honour usually ascribed to Thomas Young in 1802).[26]

One last thing, how did the Principle later get to be called the Principle of Virtual *Work*? The answer is that the Principle applies at an instant, and an instantaneous velocity is proportional to an instantaneous displacement (even while they are both 'virtual'), and the scalar product of a force and a displacement is not just 'energy', it is that special kind of energy known as 'work'. All this will be explained again in more detail in Chapter 4.

[25] Dugas R, *A History of Mechanics*, Dover Publications, Inc. New York (1988) page 233.

[26] Coopersmith, *EtSC*.

The Provenance of the Principle of Least Action: Maupertuis

Here is yet another principle in this plethora of principles, and it's the one that will turn out to be the underlying principle, incorporating all the others. The principle arose out of successive generalizations of earlier principles:

(i) first, as we have already seen, there was Heron's 'shortest path' for reflected light (second century AD),

(ii) then there was Fermat's 'Least Time' for reflection *and* refraction (1662),

(iii) then Leibniz, in 1682, in his Principle of Least Resistance, pooh poohed Fermat's Principle, for why should light make a choice between optimizing 'time' and optimizing 'distance'? No, argued Leibniz, light takes the easiest path, the one for which the 'resistance' is least,

(iv) finally, Maupertuis, in 1744, extended Leibniz's principle to cover the motion of light *and* bodies - he called it the Principle of Least Action.

The tale of Maupertuis and the Principle of Least Action is an entertaining one, redolent of the eighteenth century, and full of metaphysics, intrigue, and curious characters, and so we will let it divert us for a while. Lanczos whets our appetite with the following rousing words: "[the eighteenth century] is the only period of cosmic thinking in the entire history of Europe since the time of the Greeks."[27]

Pierre-Louis Moreau de Maupertuis, son of a wealthy pirate, was born in St Malo, France (1698–1759). Maupertuis was variously a musketeer, geographer, amateur astronomer, biologist, moralist, linguist, and surveyor. In this last capacity, Louis XV commissioned him to lead an expedition to Lapland to measure the length of a degree along the Earth's meridian; Maupertuis's findings corroborated Newton's predictions - that the Earth was flatter at the poles. Voltaire, famous French wit, and man of letters, was delighted and awarded Maupertuis the epithet 'Earth-flattener' (while also lampooning him for having brought

[27] Lanczos, Preface, p x. (Mind you, what about the cosmic laws of thermodynamics brought in by Clausius and by Thomson in the nineteenth century, and Einstein's theories in the twentieth century?)

two Lap women back to Paris). Voltaire later (around 1740) recommended Maupertuis to be President of the new Prussian Academy of Sciences being set up by Frederick the Great in Berlin (Frederick preferred French scholars, they represented the height of culture and refinement).

Around the early 1740s, Maupertuis discovered the Principle of Least Action, getting the germ of the idea, and the word 'action', from Leibniz's Principle of Least Resistance, but extending the latter to cover the motion of bodies (Leibniz's Principle considered only light). Maupertuis defined action as mvs, where m is the mass of the body, v its speed and s the path. For just one body the mass dropped away as merely a constant multiplying factor (its constancy was never in question at this stage in physics), and the quantity to be minimized was then the sum over vds, where ds was a small increment of distance travelled along the path.

The discovery of this principle was a turning point in Maupertuis's career, he vaunted it and made it a cornerstone of his philosophy. In his essay "The laws of rest and of motion deduced from the attributes of God" he wrote:

> "Whenever any change takes place in Nature, the amount of action expended in this change is always the smallest possible."[28]

Maupertuis also applied the principle to scenarios outside the purely physical realm (in his calculus of pleasure and pain, the total happiness was maximized and the total pain minimized) and, overstepping the mark (by modern standards anyway), argued that least action was proof of the existence of a "Supreme Being".[29]

The response to the principle was varied: the Leibnizians did not rate the principle as high as Leibniz's conservation of 'live force'; the atheists/materialists objected to Maupertuis's use of the principle as proof of God's existence; and some, especially the *philosophe* and encyclopaedist, d'Alembert, objected to the teleological implications - how did the body know which path was the minimum one? Just one philosopher was unequivocally on Maupertuis's side - the great Euler.

[28] de Maupertuis, P L M, *Les lois de mouvement et du repos, déduites d'un principe de métaphysique.* (1746) Mém. Ac. Berlin, p. 267.

[29] Reference in Jourdain P, The nature and validity of the Principle of Least Action, The Monist, Vol 23 (1913) page 11, note 40.

Now one Samuel König (1712–57) had recently been elected to the Academy in Berlin, at Maupertuis's invitation (König, Maupertuis, Voltaire, and Voltaire's mistress, the marchioness, Mme du Châtelet, were all friends, promoting 'Newtonianism' in France, and often staying at Voltaire's mansion at Cirey near Geneva). Nevertheless, König attacked Maupertuis's Principle, saying that, on the one hand, Maupertuis had got it from Leibniz, and on the other hand, it wasn't correct. Maupertuis demanded to see proof of Leibniz's priority, which - said König - was in some letters that Leibniz had written to Hermann (a mathematician). König could not find the originals, which, apparently, were in the possession of a certain Henzi of Berne, who had been decapitated. Maupertuis was incensed, and wrote eleven times to Hermann's heirs (via Johann Bernoulli (II) in Basle) asking them to hunt through all the old correspondence. Even so, the letters were not found (and they never have been[30]). Maupertuis then arranged for a hearing of the Academy to determine whether König's copies were forgeries. On the day (13th April 1752) Maupertuis was absent and Euler presided, but the result was a foregone conclusion (the Academicians were hardly impartial, as they relied on Maupertuis for their promotions). König was found guilty, appealed, lost again (8th June 1752), and then resigned from the Academy. Maupertuis, aware that these consequences made him unpopular, fought even more vigorously, and asked the Princess of Orange at the Hague to threaten König with dismissal from his post as Court librarian.

Then Voltaire got on the bandwagon, forgetting earlier epithets, and now criticizing Maupertuis for plagiarism, bad physics, being a tyrant, and anything else he could think of. This century saw the birth of satire and it was most unfortunate for Maupertuis to get on the wrong side of Voltaire, the *facileprinceps* of satire. To top it all, Voltaire was in a dismal mood as Mme du Châtelet had recently died in childbirth. He proceeded to write an entire book with the express purpose of pouring abuse on Maupertuis, casting him as a stupid, presumptious fellow, the student of a certain Dr Akakia ('kak' would have sounded just as offensive in the eighteenth century as it does today).

Frederick II defended Maupertuis, the president of his Academy, and Euler also came to Maupertuis's aid. In fact, Euler's defence of

[30] Instead, some relevant letters between Leibniz and various Bernoulli personae were found in the Bernoulli family archives by historian, Kabitz, in 1913.

Maupertuis's priority was very generous, especially considering that Euler had found some serious errors in Maupertuis's formulation, and had actually discovered the principle the year before Maupertuis! (Maupertuis's formulation was vague and incorrect and he needed Euler to tidy it up, changing the sum into an integral, and showing that the principle was meaningless unless the conservation of energy applied). As Lanczos writes:[31] "Although Euler must have seen the weakness of Maupertuis' argument, he refrained from any criticism, and refrained from so much as mentioning his own achievements in this field, putting all his authority in favour of proclaiming Maupertuis as the inventor of the Principle of Least Action. Even knowing Euler's extraordinarily generous and appreciative character, this self-effacing and self-denying modesty has no parallel in the entire history of science".

There has also been no parallel to Voltaire's campaign of vilification. Maupertuis's health was badly affected by the stress, and some while later he left Berlin for St Malo, still pursued by vitriolic volleys from Voltaire's pen, never to return (Maupertuis died at a stop-over at Johann (II) Bernoulli's in Basle).

The verdict of today? König was in all probability honest but naive and earnest, not appreciating that one must always allow the President of an Academy to have a face-saver. Maupertuis, although by all accounts an uppish fellow ("spoilt, intransigent,... very short and always moving, with many tics, careless of his apparel."[32]) nevertheless had genuinely stumbled into an idea of great and long-lasting importance. Although getting the initial idea from Leibniz - a source which Maupertuis didn't deny - Maupertuis had taken it to a new level of generality (by applying it to masses). It is possible that Leibniz also proceeded to this step, but no evidence in the form of any original letters or papers was found at the time or since. Maupertuis's outrage, and his attempts to track down Leibniz's letters, appear genuine. It is true that Maupertuis was not in the first rank of mathematicians (mind you, being second to Leibniz and to Euler was not bad going) but, alone amongst his contemporaries, he did intuit the principle's cosmic significance. Not even Euler had done this. Giving the last words to Euler:

[31] Lanczos, page 346.

[32] *Dictionary of Scientific Biography* (the article on Maupertuis), ed Gillispie, C C, Charles Scribner's Sons, New York (1970–80).

> "This great geometer [Maupertuis] has not only established the principle [of Least Action] more firmly than I had done, but his method, more ubiquitous and penetrating than mine, has discovered consequences that I had not obtained."[33]

and

> "nobody before the Illustrious President of our Academy [Maupertuis] has even suspected in what elements this principle was contained and how it could be accommodated to all cases. As regards myself, I only knew in a sure manner *a posteriori* the principle I used to determine trajectories; and I have ingenuously confessed that I was not in a position to establish its truth in another manner."[34]

It seems that even the greatest of mathematicians can sometimes benefit from the intuitions of a generalist.

The Variational Mechanics

This ends our brief survey of the antecedents. The Principle of Least Action proper was founded by Lagrange, and by Hamilton, with important contributions from d'Alembert, and from Jacobi. It is hardly surprising that this crowning glory of human thought should have been brought in by remarkable and unusual individuals.

Jean le Rond d'Alembert (1717–83) (the source of the quote heading Chapter 1) was a foundling, and given the name 'Jean Le Rond' from the church in Paris on whose steps he was left. The police traced his parentage to a famous *salonniere* and a cavalry officer. The *salonniere* never acknowledged her son but the Chevalier arranged for him to be fostered by a humble glazier and his wife. D'Alembert wrote his most famous works while living with his foster mother for 48 years. He finally "weaned"[35] himself after 48 years, and then lived with his mistress, herself a famous *salonniere*. As well as being a mathematician and *philosophe*, d'Alembert was also joint editor of the famous *Encyclopédie*

[33] Dugas R, *A History of Mechanics*, Dover Publications Inc. (1988) page 271.

[34] Reference in Jourdain P, The nature and validity of the Principle of Least Action, The Monist, Vol 23 (1913) pages 26–7, note 114.

[35] D'Alembert's own expression - see the article on d'Alembert in the Dictionary of Scientific Biography, ed Gillispie, C C, Charles Scribner's Sons, New York (1970–80).

with Denis Diderot (until the two of them fell out, and d'Alembert left the enterprise to Diderot). The 'd'Alembert's Principle' is explained in Chapter 5. It arises from his *Traité de Dynamique*, 1743, a work that seems to try to make every new idea as hard as possible to comprehend. (For example, it is startling for the modern mind to learn that mechanics is not empirical - according to d'Alembert, it is a purely rational subject.)

Joseph-Louis Lagrange (1736–1813), Italian by birth, was director of mathematics at the Prussian Academy of Sciences in Berlin (at the invitation of Euler and d'Alembert), then moved to revolutionary Paris in 1787, and continued his illustrious career as Professor at the École Normale (briefly), the École Polytechnique, the Bureau des Longitudes, and was a Senator in 1799. He managed to stay in favour on both sides of the French Revolution (1789) by cannily adopting his maxim: "every wise man [should] conform strictly to the rules of the country in which he is living, even if they are unreasonable."[36] He had a long happy marriage, was widowed, and then had a second long happy marriage; he expressly did not have children as they would be a distraction from work. One has the impression of a person who purposely led an uneventful life - this is often true of the most creative individuals.[37]

Lagrange discovered an outstanding method of minimizing integrals in max/min problems - the 'calculus of variations'. Also outstanding was his 'Method of Lagrange Multipliers' (see Appendix A6.4). His masterpiece was the *Mécanique analytique*,[38] published in 1788, the most comprehensive account of mechanics since Newton's *Principia* from one hundred and one years earlier, and radical in being completely general (not tied to this or that mechanical scenario). It was unusual in another way - Lagrange announced in the preface that the work contained no diagrams, or mechanical arguments, but "solely algebraic operations".[39]

William Rowan Hamilton (1805–65), from Dublin, Ireland, was a linguistic and mathematical prodigy. Aged eight, he unravelled the methods used by the American Calculating Boy, Zerah Colburn, and before his thirteenth birthday he knew thirteen languages. He later

[36] Dictionary of Scientific Biography (see earlier footnote), the article on Lagrange, p 569.

[37] Newton, for example, never got around to visiting Oxford let alone abroad - but then, he could see the world in a pebble or a raindrop.

[38] Lagrange, J-L, *Mécanique analytique* 1788, Cambridge Library Collection, Cambridge University Press, 2009.

[39] Imagine a book on mechanics today without a single diagram.

considered both poetry and mathematics as the wellsprings of his creativity (his friend, the poet William Wordsworth, tactfully told him to stick to mathematics). Due to the influence of another friend, the poet Samuel Taylor Coleridge, Hamilton was attracted by the idealist philosophy of the German philosopher, Immanuel Kant. Hamilton's idealism also extended to his love affairs: his first love represented the ideal but remained on an abstract plane (she was forced by circumstance to marry another).

Hamilton was enormously impressed by Lagrange: he likened Lagrange to Shakespeare, and the *Mécanique analytique* to a scientific poem. It was this work which inspired him to the topic of mechanics which, like Lagrange, he tried to found in the most general possible terms, and purely algebraically. In fact, Hamilton's mechanics (and optico-mechanical analogy) was so abstract and mathematically challenging that it lay unappreciated by all but one contemporary. From Hamilton's optics and mechanics he came up with just one experimental prediction - that a beam of light rays would in certain cases be refracted into a 3-D cone of light. He was lauded for this - but in the main he was simply years ahead of his time, and the dividends were reaped only in the following century. His work is explained in Chapter 7.

The German mathematician, Carl Gustav Jacob Jacobi (1804–51), was that lone contemporary who did recognize Hamilton's genius. In admiration, he coined the descriptor 'canonical' for Hamilton's Equations - a curious choice for one of Jewish parentage, but perhaps Jacobi had the enthusiasm of the convert (he converted to Christianity while a student in Berlin).[40] Readers may be familiar with Jacobi on account of 'the Jacobian' - a mathematical tool for transforming from one set of coordinates to another. Jacobi was indeed a great algebraicist, and his maxim was apparently "*man muss immer umkehren*" ("invert, always invert"). He enters our story as one who developed the 'Hamilton-Jacobi Theory' - a method for making Hamilton's Mechanics workable (in the form reached by Hamilton, the equations were mostly too difficult to solve in any actual applications).

Portraits of the natural philosophers and physicists who discovered or used the Principle of Least Action are given in Appendix A2.1.

[40] The adjective 'canonical' means 'according to canon law'. To the author's knowledge, Jacobi was the first to use this term in a secular mathematical setting.

3

Mathematics and physics preliminaries: of hills and plains and other things

3.1 Coordinates

The programme of physics is the mapping of numbers onto physical things, then the carrying out of mathematical operations using these numbers, and then, finally, the translation back to the physical things, in other words, the making of physical predictions.[1] However, there is no unique way to make the initial mapping or even to decide what needs to be mapped.

Galileo (1564–1642) was one of the first to stress that quantification was crucial - "This book [of the universe] is written in the mathematical language"[2] he said. The first and most obvious quantity to be mapped was 'length' or 'distance'. Galileo, in his famous experiments on freely-falling bodies, measured the distances travelled by those bodies (he used units such as *dito*, *canna*, and *braccia*[3]) and he realized that idealizing assumptions had to be made (about air resistance, friction, the flatness of a surface, the roundness of a ball, and so on). He also had the important realization that the different directions of 'space' were independent of each other. (For example, the horizontal and vertical distances travelled by a cannon ball could be calculated completely separately, but then could be combined to yield a parabolic path overall.)

[1] (This is paraphrased from Lanczos, page 7.)

[2] Galileo Galilei, *Opere Il Saggiatore* (The Assayer), 1623, translated by Stillman Drake in *Discoveries and Opinions of Galileo*, An Anchor Book, Doubleday (1957).

[3] *dito* is a thumb's breadth (roughly, an inch), *canna* or ell was around 39 inches, *braccia* was 21 to 22 inches. Galileo Galilei, *Dialogues concerning the two chief world systems,* (1632) translated by Stillman Drake, note on page 22.

The Lazy Universe. Jennifer Coopersmith, Oxford University Press (2017).
 DOI 10.1093/acprof:oso/9780198743040.001.0001

Galileo's view of space was affected by his ideas about motion. His outstanding insight was that all motion is relative (his Principle of Relativity). This, in turn, had some outstanding implications: that there is no absolute state of motion, that there is no absolute state of rest, and that, left to itself,[4] a body travelling horizontally with constant speed will carry on travelling at that speed forever. This sounds a bit like Newton's First Law of Motion, still fifty years into the future, but, instead of Newton's 'inertial' motion, Galileo identified two kinds of 'natural' motion: vertical accelerated motion of a freely-falling body; and horizontal motion at constant speed of a body at a fixed distance from a gravitating centre.[5] Therefore Galileo's never-ending horizontal motion was not motion in a straight line, but was motion in a *circle*. In other words, for Galileo, all 'natural' motion was connected with a gravitating centre - either free-fall towards it, or circling around it - and the idea of motion proceeding infinitely and rectilinearly into empty space, seemingly for no reason, was unthinkable.

Descartes (1596–1650) did think about it. In his view, the Universe had to be infinite in extent otherwise there would have to be a boundary and this would be counter to God's perfection. He introduced a new mathematical scheme which would change our worldview forever.[6] Space was represented by three[7] independent directions (x, y, and z 'axes'), infinite, straight, and at right angles to each other. The axes were marked off at regular intervals, and so an arbitrary point in space, P, could be identified ('coordinated') by three numbers, the *coordinates* (x, y, z). The consequences were extraordinary: in addition to the geometric proofs of the Ancient Greeks (the visual properties of straight lines, circles, triangles, and so on) there developed proofs that were purely algebraic (the equations of straight lines, circles, etc.). The techniques of this new 'coordinate geometry', meant that all the properties of parallel lines, right-angle triangles, and so on, could be explored even by a blind person, from within the space (there was no need for an extra dimension in which to view the shapes).

[4] For example, the body is not propelled by a cannon, or impeded by air resistance.

[5] Apart from these 'natural' motions, Galileo also identified 'forced' or 'violent' motions, the motion of things that were thrown, shot, or otherwise forcibly projected. A gravitating centre was a mass of planetary size.

[6] Actually, Fermat had the same idea independently of Descartes. Sometimes, when the time is ripe, then many thinkers have the same revolutionary ideas - but perhaps just two thinkers is not enough to draw conclusions in this case.

[7] Descartes had just two axes, x and y, and it was Fermat who brought in the third, z.

Descartes not only extended Galileo's space to an infinite one, he also extended Galileo's conception of motion. While still holding on to Galileo's Principle of Relativity, he made it apply to all uniform motion, whether horizontal or not, and rejected Galileo's distinction between 'violent' and 'natural' motion. Also, curiously and presciently, despite his new infinite axes, Descartes thought that only *local* motion made any sense.[8]

Newton (1643–1727), when a student, found a copy of Descartes's *La Geometrie* at a fair, and was about to throw the book away in disgust when his tutor persuaded him otherwise.[9] Descartes became one of the giants upon whose shoulders Newton stood, and Descartes's infinite space (and also his concept of 'inertial' motion) was passed on, through Newton, to succeeding generations. In 1687 Newton published his famous *Principia*[10] containing his three Laws of Motion (Appendix A1.1). The First Law relied upon a space that was universal and featureless - there were no absolute markers in it and so positions, distances, and uniform motions (that is, unaccelerated motions) could be determined only relatively (that is, relative to each other or to some imagined reference frame). By contrast, the forces and accelerations in the Second Law of Motion were absolute. To give an everyday example, a relation such as 'to-the-right-of' is not absolute - it depends on the position of your eye - whereas two magnets that are attracted accelerate towards each other and eventually collide howsoever you view them, and even howsoever you are moving as you view them. In summary, Newton's Laws required that there was one, true, absolute space, and that this space was a featureless, infinite, passive background to the events (forces and accelerations) within.[11]

One hundred and one years after Newton's *Principia* (1687), Lagrange published his *Mécanique analytique* (Analytical Mechanics) (1788), and changed the concept of 'coordinates' yet again. In Newton's worldview, particles and forces are dropped into space, an infinite Cartesian

[8] Westfall R, *Force in Newton's Physics* (1971) American Elsevier and Macdonald, London, page 59.

[9] Westfall R, *Never at Rest: a biography of Isaac Newton* (1983) Cambridge University Press.

[10] Newton I, *The Mathematical Principles of Natural Philosophy* (1687) translated 1729 by Andrew Motte.

[11] Curiously, Newton's one, absolute space can be represented by an infinity of allowed ('inertial') reference frames - but it is to be understood that these are all equivalent to each other (they are identified in the next footnote).

'reference frame'.[12] Lagrange, however, uses coordinates that are tailored to the given scenario, and that must be reformulated for every new scenario. For example, consider going from a system with just one particle to one comprising N particles. According to Lagrange, the new system has $3N$ coordinates, $(x_1, y_1, z_1, x_2, y_2, z_2, \ldots. x_N, y_N, z_N)$, and therefore $3N$ dimensions of 'space' whereas, according to Newton, there are always just 3 space dimensions (x, y, and z) whatever the number of particles. The fundamental difference is that, in Lagrange's mechanics, particles are not dropped *into* space, they *are* space.

Lagrange went even further: not just particle-positions but *any* continuously graduated and quantifiable physical attributes can count as coordinates. For example, we can map the angle, θ, that a simple pendulum swings through; or the distance, s, that a bead travels as it moves along a curved wire; or the radial distance, r, and angles, θ and ϕ, of an orbiting satellite; and so on. So far, these coordinates are all distances or angles - generalizing yet further, we can assign coordinates to other properties: the changing voltage between capacitor plates, the area of a soap bubble, the values of the coefficients in the Fourier expansion of a waveform, and so on. In other words, any physical property which varies continuously, and which characterizes the system may be a coordinate of that system. Lagrange's coordinates are called the *generalized coordinates* and are designated q_i, with i running from 1 to n, and where n is as large as necessary in the given scenario. (Notation: we will write $\{q_i\}$ as a shorthand for the set of coordinates $q_1, q_2, q_3, \ldots q_n$.)

Isn't this a retrogade step, to move from the breathtaking abstraction of an infinite, eternal, empty space to a set of coordinates that are system-specific and only as extensive as the system requires? Well, it might have been a backwards step if it wasn't for one surprising and outstanding advantage: the new Lagrangian formulation allows us to sacrifice a universal space in favour of a universal physical principle. Before we explain this remarkable progression, we must attend to another 'coordinate' that we have forgotten till now.

3.2 Time

According to historian of science, Charles Gillispie (1918–2015), Galileo was the one who really brought 'time' into science. Yes, the seven ages

[12] The reference frames may be in different positions, orientations, or have different states of uniform motion.

of man were known about, yes, the acorn took time to grow into an oak, and, yes, time could be marked out by water-clocks, candle-clocks, and sundials: but that time could be put into a *mathematical* relationship, that it could be brought into comparison with distances travelled - this was new.[13] This was such a difficult step to take that it took Galileo over twenty years to be able to state his law of free-fall as 'the distance fallen is proportional to the squares of the *time*'.[14]

The next big advance came with Newton. He brought 'time' into dynamics (via his Second Law of Motion, $\mathbf{a} = \mathbf{F}/m$, where acceleration is the rate of change of velocity with *time*), but, as m and $\mathbf{F}$ could be absolutely determined, then this implied that time was absolute also. As Newton wrote in the *Principia*: "Absolute, true and mathematical time, of itself, and from its own nature, flows equably without relation to anything external...".[15]

'Time' can be quantified, and it varies continuously, so is it to be counted as a generalized coordinate? The answer is both yes and no. 'Time' is special in that it is the obvious choice as the freely chosen or independent coordinate, the one against which all the others are measured (for example, speed is distance travelled in a given time, acceleration is the velocity change in a given time, and so on). Yet, there is no logical necessity for 'time' to take on this role, and no logical prohibition against another coordinate being the independent one (we can have a standing wave plotted against horizontal distance, the height of an embankment against the speed of river-flow, and other examples). 'Time' is also special in that there is only one dimension of time - but, again, there is no logical prohibition against time having two or more dimensions.[16] It seems that we must simply accept that, empirically as well as otherwise, 'time' (designated t) is different from the other coordinates (designated q_i, and sometimes called the generalized *position* coordinates in contradistinction to the time coordinate). As Feynman says in his *Lectures on Physics*,[17] we must simply accept that time is, after all, not the same as space.

[13] Although there was already astronomical time, for example, in Ptolemy's epicycles, and Kepler's Third Law of planetary orbits.

[14] Gillispie C C, *The Edge of Objectivity*, Princeton University Press (1973).

[15] *Scholium* to the Definitions in the *Principia* (see earlier footnote), Bk I (1689) transl. Andrew Motte (1729), Florian Cajori, Berkeley, University of California Press (1934) page 6.

[16] (Two time dimensions were, in fact, proposed by Dirac and by Milne in the 1930s.)

[17] The quote is somewhere in the *Feynman Lectures on Physics*, probably Volume I, but - my apologies - I haven't been able to find it again.

The mechanics of both Newton and Lagrange splits into two regimes - statics (there is no time-dependence) and dynamics (there is a time-dependence). However, Lagrange was the first physicist whose dynamics, on occasion, treated 'time' on a more equal footing with the other coordinates. He sometimes lumped all the coordinates, including time, together, making them all dependent on some new 'dummy variable'. Time is special in yet another regard. When all but microscopic phenomena are investigated, time appears to flow, and always in one direction (from 'the past' to 'the future'). It's very strange to say, but this profound yet banal human experience of time plays no part whatsoever in the dynamics of either Newton or Lagrange. Even though the dynamics examines macroscopic effects (but, crucially, microscopic dissipative effects, like friction or air resistance,[18] are ignored), there is no sense of time flowing, no difference between making time run forwards or backwards in the equations. As Einstein wrote: "...the distinction between past, present, and future is only an illusion, however persistent."[19]

3.3 Degrees of Freedom

We have our coordinates, say, n of them for the given system. There may still be different coordinate representations, connected by transformation functions, for this given system (for example, we may use Cartesian, or spherical polar coordinates). Also, there may be extra conditions - such as the conditions of rigidity, incompressibility, or inextensibility. Say there are m such conditions. Then there will be only $n - m$ truly independent coordinates. Now when we have explored lots of different coordinate representations, then that one using the smallest number of independent coordinates determines the 'number of degrees of freedom' of the given system. However, it must be admitted that the 'smallest number of independent coordinates' is a dry and overly formalistic criterion for something so important. Physically, the degrees of freedom are *the* defining characteristics of the given system - the irreducible, independent 'motions' of which the system is capable - and their explanatory power is just as important as the economy of

[18] But Newton *did* treat the case of motion through a resistive medium in the *Principia*, Book II.

[19] Albert Einstein and Michel Besso: *Correspondence 1903–1955*, p. Speziali, ed Paris, Hermann 1972, as found in Brian Greene's *The Fabric of the Cosmos*, p 139, Penguin, 2008.

coordinates. While the degrees-of-freedom is ultimately a somewhat slippery concept, it is, nevertheless, an inherent property of the given system, and not merely a property of this or that coordinate description.

Economy may be a good thing, but how can we be sure that we have *enough* coordinates to completely describe the system? This is really the same as asking how we can be sure we have taken account of all the effects. In this regard Einstein, as so often, comes to our aid. Suppose, for example, that we are monitoring waves breaking on the beach. The waves arrive every few seconds and have a wavelength of a metre or so, but then we realize that we forgot to take account of the swell with a wavelength of one or two kilometres. This swell was not noticeable on our length and time scales, and - this is Einstein's message - if an effect isn't noticeable then we don't need to notice it.[20] In other words, if we think we have enough q_i to describe the system, then we probably do.

3.4 A generalized mechanics

We have gone from Cartesian coordinates to generalized coordinates, and from Newton's particles to 'generalized particles' (lever arm, pendulum bob, capacitor plates, and so on). Force will become generalized as well, but it is part of a whole change of tack. We move away from Newton's outstanding and simple '$\mathbf{F} = m\mathbf{a}$', to a new *energy* analysis in which 'force' is replaced by the '*work* done by a force'. We remember from high-school physics that the work done by a force, $\mathbf{F}$, acting through an infinitesimal[21] displacement, $d\mathbf{r}$, is given by the scalar product $\mathbf{F} \cdot d\mathbf{r}$. For example, for N Newtonian particles with position vectors, $\mathbf{r}_i$, displacements, $d\mathbf{r}_i$, and each subject to a force, $\mathbf{F}_i$, then the total work done is:

$$dW = \mathbf{F}_1 \cdot d\mathbf{r}_1 + \ldots + \mathbf{F}_i \cdot d\mathbf{r}_i + \ldots + \mathbf{F}_N \cdot d\mathbf{r}_N \qquad (3.1)$$

What is rarely stressed is that this is a highly specialized definition in which $\mathbf{F}_i$ and $d\mathbf{r}_i$ are '*rectangular* vectors', that is, they are referred to the

[20] (This is connected with the precision in the detection apparatus. Note, also, that this pre-dates any quantum mechanical considerations about the precision of measurements, the rôle of the observer, and so on.)

[21] Later, Section 3.6, we explain that our test of stationarity involves *differential* geometry - that's why we are using infinitesimal *differentials* rather than finite displacements.

rectangular[22] axes, x, y, and z. Making the switch to the generalized coordinates, we have the generalized work:

$$dW = Q_1 \cdot dq_1 + \ldots + Q_i \cdot dq_i + \ldots + Q_{3N} \cdot dq_{3N} \tag{3.2}$$

where dq_i are infinitesimal changes in the coordinates, q_i, and the Q_i are functions of these coordinates, $Q_i = Q_i(q_1, q_2, \ldots q_{3N})$. Comparison of equations (3.1) and (3.2) suggests that the 'Q_i' may be regarded as stand-ins for the '$\mathbf{F}_i$'. Yes, but note several things: the i in equations (3.1) and (3.2) have different ranges; the $\mathbf{F}_i$ is a vector and is the total force (including constraint- and internal-forces) on the ith Newtonian particle; 'Q_i' is not necessarily a vector,[23] and applies to a 'generalized particle' as it goes through a 'generalized displacement', dq_i. This may all sound like mere formalistic pedantry, but our message will be that the formalism (the mathematics) has *physical* implications (and vice versa). For example, a 'generalized particle', by its nature, moves only in a certain way, in harmony with the constraints and kinematic conditions (a car travels along a road, but a 4-wheel-drive can go over the rough, while a person walks along a footpath[24]). As the 'generalized particle' moves harmoniously, that is, in keeping with the constraints and kinematic conditions, then it doesn't do work against these constraints and kinematic conditions. Therefore the generalized forces, Q_i, do not include constraints and kinematic conditions - they represent applied or external effects only. This is a major difference between the forces and the generalized forces.

More to do with formalism: the total work in equation (3.2), sometimes symbolized U, is what, in the old classical physics, used to be called the 'work function'. The term 'function' is used advisedly. 'Work' is a form of energy, yes, but it is a rather special kind of energy, having a *functional* dependence on the distribution of various macroscopic components (like the raising of weights, the stretching of a spring, and so on). Remembering also that there is a functional connection between the different coordinate representations (Section 3.3), then it is hardly surprising that the $Q_1, Q_2, \ldots$ end up as being functions of the $q_1, q_2, \ldots$ They

[22] Rectangular axes are straight (rectilinear) and at right angles to each other (orthogonal).

[23] In fact, Q_i and dq_i are more general 'vectors' that exist in an abstract 'vector space'.

[24] Other examples are: a lever arm swings about the fulcrum, a spring stretches along its axis, a rigid body moves while still maintaining its shape because of internal forces.

may also be functions of t. The $Q_1, Q_2, \ldots$ are known as the generalized forces. As already mentioned, they are not necessarily rectangular vectors, and do not necessarily have the dimensions of force. The only requirement is that, while the generalized displacements need not have the dimensions of length, and the generalized forces need not have the dimensions of force, each product, $Q_i dq_i$, *must* have the dimensions of energy.

Although it seems that the work function must be less fundamental than the force (as it requires force in its definition), this mistaken hierarchy is partly due to historical accident (force was discovered first), and partly due to the fact that the force-analysis is simpler and more intuitive. But, in fact, it is the work done that is more fundamental. As Lanczos writes:[25] "Although we are inclined to believe that force is something primitive and irreducible, the [variational] mechanics shows that it is not the force but the *work* done by the force which is of primary importance..." Now in the Newtonian Mechanics a force causes a particle to accelerate whereas in the Variational Mechanics the 'work done' causes a 'generalized particle' to change its kinetic energy. This is why Lanczos also writes: "the really fundamental quantity ... is not the [rate of change of] momentum but the kinetic energy."[26] 'Work done' and 'kinetic energy' are both forms of *energy*, and this confirms our starting assertion that Newton's 'force-mechanics' is being replaced by a new 'energy-mechanics'.

We shall also find that the dichotomy is not just between force and energy, but between a *whole-system* feature (the work function or configuration-energy) and the *individual* energies (the kinetic energy of each individual mobile component). Now a system is evidently made up of its components, but the boundary between 'whole' and 'parts' is not always clear-cut. In the variational mechanics the motion of components, and even the mass of an individual component, can affect the 'whole', whereas in Newtonian Mechanics the influence between '$\mathbf{F}$' and '$m\mathbf{a}$' is all one way.

Curiously, while energy is in the ascendant in the Variational Mechanics, the well-known Principle of the Conservation of Energy will be relegated to a subordinate role, applying only in the special case where there is no explicit dependence on time. In the special case in which the work function, U, is time-independent and also doesn't

[25] Lanczos, page 27.

[26] Lanczos, pages 21–2. We explain inertia in Section 3.5.

depend on the 'speeds', $\dot{q}_1, \dot{q}_2, \ldots,$[27] then U is the same as the potential energy, V (actually, by convention, $U = -V$).

3.5 'Space' - a mathematical testing-ground

We have generalized coordinates, generalized particles, and generalized forces; we also have need of a new generalized conception of 'space'. A crucial requirement is that our choice of 'space' - a mere viewpoint - must not make any difference to the system being investigated. This requirement is certainly satisfied when switching between, say, rectilinear and curvilinear[28] coordinates - this is a purely mathematical procedure effected by mathematical transformations between coordinates. However, we can go from one space,[29] $\{q_i\}$, to a second space, $\{q'_j\}$, in another way - we can adopt two completely different *physical* models for the one given system (for example, we can model bells, or their clangers).

However, space and physical assumptions are intimately related. Let's examine this relationship first in the familiar territory of Newtonian Mechanics. By Newton's First Law of Motion, a free particle travels in a straight line at constant speed - but how shall we know that this particle, coasting along in empty (Newtonian) Space, is in fact moving through regular distance intervals in regular Time intervals, in other words, how shall we know that it is really moving in a *straight* line? We can only be assured of this by postulate - that there really are no forces operating, and that the universality and regularity of our rulers and clocks can be relied upon. How can we be assured that our free particle is not swerving to the left? Again, we can only be assured of this by postulate. Also, if the particle is not free but *is* subject to a force then it will have an acceleration determined by that force, and also by the particle's mass. How can we be sure of this? Because we assert that our reference frame is definitively not accelerating, and also that the mass is constant between reference frames.

[27] Notation: a dot over a symbol is a shorthand for d/dt.

[28] Note that curvilinear coordinates, such as (r, θ, ϕ), do not make a flat 'space' become 'curved'. The true meaning of 'curved' will be explained in Section 3.6.

[29] Notation: we mostly drop apostrophes round the word space from now on. The curly brackets, $\{q_i\}$, is a shorthand for the whole set of n coordinates: $q_1, q_2, \ldots, q_n$. The prime, q'_j, means a different space, it does not mean differentiation.

In the Variational Mechanics, a 'free' generalized particle is one that goes through its paces, its degrees of *freedom*, unhindered: the pendulum bob swings, a marble rolls down the marble-run, the Earth follows an elliptical orbit around the Sun, a pebble falls freely until the instant before it reaches the ground.[30] So, 'free particles' can swerve, speed up, or slow down, - all accelerations - but accelerations can now be considered as 'inertial', and sometimes even define the new 'straight' lines of space.[31] In effect, the concept of 'inertial' has also been generalized, but the new 'inertial' is no longer divorced from 'space'. For example, 'inertia' can depend on the *spatial* distribution of mass (the 'moment of inertia' is often proportional to r^2); and the response of an electrical circuit is dependent on the distribution in *space* of its components (merely curving a straight wire into a coil turns it into a new component - an electromagnet). What is perhaps less well appreciated is that time also affects space (the rate at which a bar magnet is pushed into or out of an electromagnet affects the strength of the induced currents; triangulation using lasers relies upon the regularity of clocks; and so on). Also, it is well attested that the presence of very large gravitating centres affects the surrounding space.[32]

This interrelation between space and the given physical features makes the challenge of finding a fiducial viewpoint - the challenge of objectivity - seem insurmountable. But this challenge can be met, and what we require is a space that can serve as an agreed mathematical testing-ground - a mathematical-test-*space*. How we meet this requirement is rather clever: instead of using *physical* space, we construct a completely *abstract* space. We take the n generalized coordinates for the given system, $q_1, q_2, \ldots, q_n$, and plot them against n rectangular axes (straight lines meeting at right angles) - it's as simple as that. And, in the same way as we have been used to associating the position of a point in *physical* space with the three coordinates, x, y, and z, we now associate the position of the system in *abstract* configuration space with the n coordinates $q_1, q_2, \ldots, q_n$. The 'position of the system' is known as the

[30] Putting one's finger in the way of the pebble it will no longer be free. Likewise, friction, air resistance, or booster rockets all stop it from being free.

[31] These lines are called geodesics, but this terminology only applies in time-independent cases, for example, where the external force-fields do not vary with time.

[32] See, for example, Will, Clifford, "*Was Einstein Right?: putting general relativity to the test*", Basic Books, New York, 1986.

configuration point or C-point, and it represents the state of the whole system (the values of all the q_i s) at a given time. Finally, in the same way as we can track the actual motion of a particle along a trajectory as its position changes in time, so we can track the abstract motion of the C-point as it moves along a curve, its world-line, through time.

To emphasize that this configuration space is truly abstract, we consider an example from outside the realm of physics. Imagine a two-dimensional configuration space in which we plot the number of white chess pieces against the number of black chess pieces, as a given chess game procedes. If the end-game is long and drawn out, with no pieces taken, then the C-point will remain fixed, even while the actual, physical disposition of the chess pieces keeps changing.

We now have our mathematical testing ground - configuration space - but what shall the mathematical test be? From the title of this book, we can guess that it will be a test of stationary or least action, but we must creep up on this more slowly.

3.6 Invariants: 'space research'

The layperson's perception of Einstein's legacy is that he found that 'everything is relative'. This couldn't be more wrong: the true content of Einstein's Principle of Relativity is that the important things are not relative, they are invariant (invariant with respect to different viewpoints, cf. d'Alembert's quote at the start of Chapter 1).

Let's compare two views - spaces - of one physical system. There is an important proviso, crucial to Variational Mechanics: the two spaces must be connected by coordinate transformation *functions*. In Figure 3.1 we see a column in the mirror-surface, and the same column in the table-top surface. We find that straight lines have become transformed into curves, and distances and angles are not preserved (for example, the grooves in the column are parallel in one space and diverging in the other space).

Nevertheless, there are certain geometric features that do remain the same between the two spaces: there is just one column, it always has exactly 10 straight grooves, and the leaves, whorls, and flowers are always at the top of the column (not half-way up). These are so-called topological features, and their importance lies in the fact that they are *invariant* between spaces. If our mechanics can be couched exclusively in terms of such invariant topological features then it has a chance of being universal.

Figure 3.1 "*Anamorphic Column*" by Istvan Orosz ©Istvan Orosz.

What is the topological feature we use when it comes to mechanics? Comparing the spaces in Figure 3.1, two things can be noted. One is that the position of a point is not a very robust quantity, it can change from one space to another - for example, for a reference frame whose origin is in the middle of a flower, the petals are all the way around this origin; whereas for an origin sited on the tip of one petal, then all the other petals are to one side. The other noteworthy thing is that the smaller the portion of space that is investigated, the more closely the different mappings approach each other, topologically-speaking (curved line-segments approach straightness, line-segments that diverge in the large appear parallel in the small, and so on). So, better than the position of *one* point, P, is the *difference* in position of *two* points, P and Q; and better still than the difference in position of P and Q is the difference in position of P and Q as Q gets *infinitesimally close* to P.

We are familiar with using the ordinary differential calculus to determine the tangent line at a point, P, on a curve - we examine the slope as a nearby point gets arbitrarily close to P. In configuration space we are now in a space of n dimensions but we can analogously identify a space of $(n-1)$ dimensions that is 'tangent to' the n-dimensional space at the point P. The branch of mathematics that deals with the rate of change of a 'hyper-surface'[33] at a point is called differential geometry, and it is just what's required in order to arrive at a topological feature that will be invariant. (The old-fashioned name, the absolute calculus, is better at reminding us of this asset.) We are about to find out that the topological feature that will be relevant in mechanics is the 'stationary point' (see Section 3.7) of the hyper-surface corresponding to the given configuration space.

Before returning to this discussion, let's pause for a brief tour of the history of 'space research'. We showed earlier (Section 3.5) that there was a link between the mechanical assumptions and the properties (the geometry) of space. Now it so happened that the predictions of Newton's Mechanics and the geometry of Euclid's space were so perfectly matched that (almost) everyone simply took it for granted that space - everyday physical space - was Euclidean.[34] It came as something of a shock, therefore, when Einstein declared (in 1915, in his Theory of General Relativity) that physical space was not Euclidean. For example, Einstein predicted that it wasn't true that the angles of a triangle would always sum to 180° no matter where the triangle was located, or that parallel lines remained parallel however far they were extended.

Actually, mathematicians in the nineteenth century had already discovered the possibility of non-Euclidean spaces but few realized that these abstract speculations could have any relevance to the real physical world. One who did realize this was the mathematical genius, Carl Friedrich Gauss (1777–1855). He tried to measure the angle-sum of a triangle defined by three mountain-tops - but the experiment was too crude to bring out the tiny deviation from 180°. However Gauss made an outstanding theoretical discovery: the properties of a space could be determined algebraically, that is, from totally *within* that space (in Section 3.1 we remarked that Descartes's coordinate geometry had likewise liberated us from the need to look at space from a higher dimension outside the space). Then there was Bernhard Riemann (1826–1866), who took Gauss's investigations to a new level of generality (see below). (There were others, Bolyai, and Lobachevsky, but we won't pursue them.)

[33] A hyper-surface is a surface in more than two dimensions.

[34] Euclidean means having the properties given by, and derived from, Euclid's axioms in his book *"The Elements"* (c. 300 BCE). For example, a Euclidean space is one in which the sum of the angles of a triangle always comes to 180°; parallel lines remain parallel to infinity; the circumference of any circle is 2π(radius); and so on.

Are the quintessential properties of a space determined solely by certain axioms concerning triangles, circles, parallel lines, and so on? Yes, the properties of a space are so-determined, but Riemann's brilliant discovery was that the properties of space can be determined by examining just *one formula*[35] - the formula for the infinitesimal 'distance' between neighbouring points, *ds*. Before Euclid, another ancient Greek had already come up with such a formula. This was Pythagoras (or, strictly speaking, the Pythagorean School), around 500 BCE, who promoted the famous formula known as Pythagoras's Theorem:[36]

$$\text{Pythagoras's Theorem, } (ds)^2 = (dx)^2 + (dy)^2 \tag{3.3}$$

Riemann realized some fundamental things: (i) Pythagoras's Theorem not only told of the properties of right-angled triangles, but the function (ds) (the positive square root of $(ds)^2$), was the 'distance' between two points; (ii) more than that, it was the *shortest* 'distance' between two points; (iii) most amazing of all, Pythagoras's $(ds)^2$ was but one possibility - more generally, it could be written as:

$$(ds)^2 = g_{xx}(dx)(dx) + g_{xy}(dx)(dy) + g_{yx}(dy)(dx) + g_{yy}(dy)(dy) \tag{3.4}$$

where the g_{ij} are coefficients to be determined. In Pythagoras's Theorem, it just so happens that we have the values $g_{xx} = 1$, $g_{xy} = 0$, $g_{yx} = 0$, and $g_{yy} = 1$. In other cases (other spaces), the g_{ij} need not have these particular values, and need not be constants - they could even be functions of x and y.[37] This discussion has, so far, been in a space of only two dimensions, but Riemann generalized his 'distance function' still further to n dimensions:

$$\text{Riemann's distance function, } (ds)^2 = \sum_{i,j=1}^{n} g_{ij}(dq_i)(dq_j) \tag{3.5}$$

Most remarkable of all, Riemann showed that *all* the geometric properties of an n-dimensional space can be completely determined by just

[35] Moreover, it's a formula which can be determined from *within* the space.

[36] We have adapted Pythagoras's Theorem by writing it in differential form, $(ds)^2 = \ldots$, rather than in incremental form, $(\Delta s)^2 = \ldots$

[37] (with the proviso that the g_{ij} must be continuous, and twice-differentiable functions).

this 'distance', *ds* (the square root of the above equation). For example, if any of the coefficients, g_{ij}, are not constant but are functions of the coordinates, then the corresponding space is not 'flat' (Euclidean) but '*curved*'. This, finally, is what 'curved' means: it is a measure of the departure from Euclidean space, and is determined by certain functions (Riemann's 'curvature functions') of the g_{ij}-coefficients. While the 'distance' is specific to the given space, its value is invariant as regards which coordinate representation has been adopted.

In view of our quest for invariant properties, Riemann's distance function is obviously very important. In effect, Riemann's distance function enables us to turn 'space research' on its head: instead of starting with Pythagoras's Theorem (leading to a Euclidean space) we can instead *start* with a postulate of a certain distance function, *ds*, and then explore the consequential geometry of that space. (Any space characterized by a certain distance function, *ds*, will be called a Riemannian space.)

> Optional: Some well-known cases are: (i) Euclidean space, $ds = \sqrt{[(dx)^2 + (dy)^2 + (dz)^2]}$, (ii) Fermat's Principle, $ds =$ time-of-flight along a light ray, (iii) Special Relativity, $ds \equiv$ 'spacetime interval' $= \sqrt{[(dx)^2 + (dy)^2 + (dz)^2] - [c^2(dt)^2]}$, (iv) $ds = \sqrt{2T}dt$ where T is the kinetic energy in a conservative system with no potential energy, (v) $ds = \sqrt{2(E - V)}dt$ where $(E - V)$ is the kinetic energy in a conservative system with potential energy V and total energy E, (vi) General Relativity, $ds \equiv d\tau =$ an interval of 'proper time' (in units where $c = 1$).

How does all this relate to 'mechanics' (pendula swinging, juggling clubs twirling, cantilever bridges, *LC* circuits, planets orbiting, starlight passing near the Sun, and so on)? All the definitions of *ds* just given (optional reading) are examples of special spaces for special cases. However in general mechanics we have bodies with mass, a potential energy function V, and V may be time-varying, and there may be additional constraints and kinematic conditions: in these most general cases, *ds is* an interval of *action*.[38]

[38] *ds* is sometimes called the 'metric' of the given space as it is an absolute measure, and gives the actual 'distance' between two specific points, for example, the points may be separated by a *ds* of 0.056 Js.

The function *ds* or 'metric' appropriate to our problem determines the 'curvature' of the space of that problem: and any mechanical problem (not just Einstein's Gravitation) may be said to occur in a *curved* space. For example, a bead travelling along a bendy wire needs more force to move it through tighter twists - it's space is not uniform, it is curved. Now a mechanical system with n degrees of freedom can be depicted using an n-dimensional configuration space, but if there are m conditions (the condition of moving along a wire, of maintaining the shape of a spinning top, of hanging on an inextensible cord, and so on) then the configuration space is reduced to a subspace of $(n-m)$ dimensions. This subspace will usually[39] be curved. For a problem with no potential energy and no explicit time-dependence, the C-point moves along a 'geodesic' - the 'straightest'/'shortest' path between given end-points - in this curved subspace. For problems with an external influence (there is a V) we can still establish what is the 'geodesic', but not determine the C-point's speed along it. However, this extra information can be reconstructed afterwards (using the condition of energy conservation). Finally, even if the external conditions are time-varying, we can add the time to the position coordinates, thereby increasing the dimensionality of the space by 1, and, again, determine the 'geodesics' of this new space. All told, *the mechanical problem has been translated into a problem of (differential) geometry*.

(It will not be possible to absorb these difficult new ideas in one go, their meaning will become more and more apparent as we proceed through the book.)

3.7 The Calculus of Variations

In this section we interrupt our story (it will be continued in Chapters 4 to 7) in order to explain some mathematical techniques. These techniques will be very helpful in the understanding of everything that follows - but they may be skipped over without too much loss of continuity.

3.7.1 Extremum problems - of hills and plains

Popularly, it is thought that hills and valleys are the antithesis of plains but actually hill-tops, valley-bottoms, and plains all have something in

[39] The subspace will be flat only in the special case where all the conditions are linear in all the coordinates. The subspace is then a 'hyper-plane'.

common. In extremum problems we are not trying to find the highest peak in a whole mountain range, but rather we are already on a given mountain and want to determine the position of the very top. Wherever we are on the mountain we ask ourselves - are we at the top yet, or does our next step take us even higher? In other words, the location of the highest point requires an inspection of the immediate - the *local* - surroundings. It is with respect to their local surroundings that hill-tops, valley-bottoms, and perfect plains are all the same - they are all *flat*. The extremum point of interest involves 'no change' relative to local, neighbouring points, and so it is called the 'stationary point' (it is said to exhibit 'stationarity').

In the variational mechanics, we still explore a 'surface', but it may have more than just the two dimensions of hills or plains (an n-dimensional surface is called a 'hyper-surface' - as mentioned in an earlier footnote). Also, we are usually only interested in whether the conditions for stationarity have been met - whether the point in question happens to be a true extremum (a maximum or a minimum), or merely an inflection point, or a point within an extended flat plain, is irrelevant.

3.7.2 *Three kinds of infinitesimal*

Suppose we have a function, y, of n variables, $x_1, x_2, \ldots, x_n$, that describes an n-dimensional space, a hyper-surface, $y = y(x_1, \ldots x_n)$. We wish to determine whether the point P in this surface is stationary or not. As we have just seen, in order to establish whether a point is stationary, we must examine the region infinitesimally close by. However, there are three distinct ways in which to take an infinitesimal step.

(1) Actual displacements

We can explore from P to a nearby position, P', by allowing the variables $x_1, x_2, \ldots, x_n$ at P, to be displaced (change their values) by very small amounts, $dx_1, dx_2, \ldots, dx_n$. You could say that we've put out tiny feelers in all the relevant directions. Then we draw the feelers back in, and 'in the limit' as P' reaches P, all the displacements are once again zero. This 'limiting process' is *mathematical* and not haphazard - the displacements scale down *in unison*, and must at every stage satisfy the function, $y(x_1, \ldots x_n)$, and must at every stage maintain the proper proportions to each other as determined by any constraints.[40] Only in this way can it be guaranteed that all the displacements will scale down and reach zero together. This mathematical down-sizing process implies that the difference between $y(x_1 + dx_1, x_2 + dx_2, \ldots, x_n + dx_n)$ at P', and $y(x_1, x_2, \ldots, x_n)$ at P, called the 'total differential', dy, is given by the well-known rules of differential calculus:

$$dy = \frac{\partial y}{\partial x_1}dx_1 + \frac{\partial y}{\partial x_2}dx_2 + \ldots + \frac{\partial y}{\partial x_n}dx_n \qquad (3.6)$$

The condition for the stationarity of y at P is that dy at P is zero.

For simplicity, let us consider the one-dimensional case in which y is a function of just one variable, $y = y(x)$. The n-dimensional 'surface' then reduces to the curve, $y(x)$. (We assume that the functional form of y is known, and that it is continuous and differentiable.) The condition for stationarity at P then becomes: $dy/dx = 0$ at P. For example, $y(x)$ could represent the vertical position of a cannon ball as a function of time; or the vertical position of a cannon ball as a function of its horizontal position; or the force on a spring as a function of the extension of the spring; or the torque on a lever-arm as a function of the angular displacement at the pivot; and so on. Let's examine the second example, the path of the cannon ball, shown in Figure 3.2.

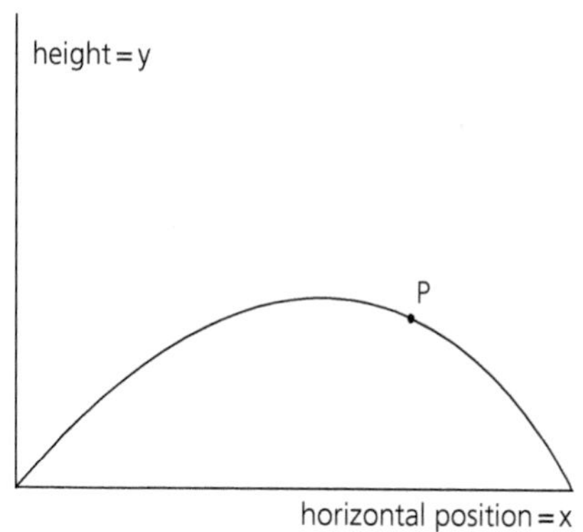

Figure 3.2 The path of a cannon ball' (schematic).

[40] (In the language of vector spaces, each displacement, dx_i, maintains a fixed direction along an axis of the abstract space, and only its magnitude changes as zero is approached.)

It so happens that in this problem we know that y is the function $y = -ax^2 + bx$ where a and b are constants. Suppose we want to determine whether y is stationary at the point P shown in the figure. There are two perspectives on this. From the physical, *experimental*, perspective, we can take successive measurements of the ball's position, plot the graph, and see whether or not the curve appears flat at P. From a purely mathematical perspective, we can use the well-known method of 'calculating the differential', dy/dx, and then check numerically and see if dy/dx is indeed equal to zero at P. (In the example shown, $y = -ax^2 + bx$, so $dy/dx = -2ax + b$, and this is equal to zero whenever $x = b/(2a)$. So, we must examine whether x at P has the value $b/(2a)$.)

Are these two perspectives, the physical and the mathematical, equivalent, leading to the same result? Amazingly, they are. It must be admitted that it's impossible to make measurements that really do get smaller and smaller all the way to zero, yet, the important thing is that the cannon ball really does undergo *actual* horizontal displacements (dx) as it whizzes through the sky, and there is, in principle, no objection to our physically sampling the curve anywhere we like.[41] (Note that, from either perspective, we just ascertain the flatness, we don't determine the height of the ball at P.)

(2) Virtual displacements

Now, we consider a different kind of infinitesimal displacement called a *virtual* displacement, δx (or, in n dimensions, the virtual displacements, $\delta x_1, \delta x_2, \ldots, \delta x_n$). To distinguish it from the previous kind of displacement we give it a different symbol, δ. We again wish to carry out a test of stationarity at the point P, and this again involves an investigation of the neighbourhood infinitesimally close by; but now, for various reasons, the displacements, $\delta x_1, \delta x_2, \ldots, \delta x_n$, are not all *physically* possible. For example, they may imply bodies having infinite speeds, or displacements of physical components which can't actually be moved (like the blocks in a Roman arch), and so on. We don't let this stand in our way, but press right ahead and mathematically imagine the displacements instead. This was Lagrange's brilliant idea (one of many).

Nevertheless, our mathematical imagination must be constrained. The new positions, $(x_i + \delta x_i)$, must still satisfy the function y (it is, after

[41] However we can only do this to *within a certain experimental precision.*

all, this same 'surface' that we are exploring) and there may also be constraints between the displacements (this is just another way of saying that the variables may not all be independent of each other). In fact, we have exactly the same mathematical requirements as we had before, and so we can now write down exactly the same equation as before, but substituting δx for dx:

$$\delta y = \frac{\partial y}{\partial x_1}\delta x_1 + \frac{\partial y}{\partial x_2}\delta x_2 + \ldots + \frac{\partial y}{\partial x_n}\delta x_n \qquad (3.7)$$

Once again, the stationarity of y at P requires the zero-value of δy at P.

Thus, in case (1) we had both a mathematical and a physical perspective, whereas now, case (2), we retain the mathematical perspective but let the requirement for an actual physical perspective fall away.

Let's consider a one-dimensional example. Suppose we have a ladder, of given mass, leaning against a wall; the wall is assumed completely smooth, but there is some contact friction between the ladder and the floor. We wish to determine the conditions for stationarity, that is, for static equilibrium at some position, P, of the foot of the ladder. Instead of balancing forces and/or balancing torques (the Newtonian method), we adopt a radically different procedure, and allow the foot of the ladder to be virtually displaced, very slightly, towards or away[42] from the wall (Figure 3.3).

However, this displacement doesn't *actually* happen, the foot of the ladder does not really slip through position P - it's not a 'falling-down-ladder', it's a 'stable-ladder' that we're investigating. By contrast, the cannon ball really does fly through position P in Figure 3.2. Thus, the infinitesimal mathematical displacements of the cannon ball are also *actual* physical displacements, whereas the infinitesimal displacements of the ladder are purely *virtual* displacements - they don't *actually* happen, they are *mathematically imagined.*

(The problems in Figures 3.2 and 3.3 are worked through in Appendix A6.1 (problem 6), and Section 4.10, Chapter 4.)

There is just one more requirement to note about the virtual displacements - they must be *reversible*, that is to say, for any displacement, δx_i, a displacement in the reverse direction, $-\delta x_i$, must also be possible. This is to ensure that we can always approach the C-point from either

[42] The reason for insisting on these *two* directions is given in the main text, the paragraph after next.

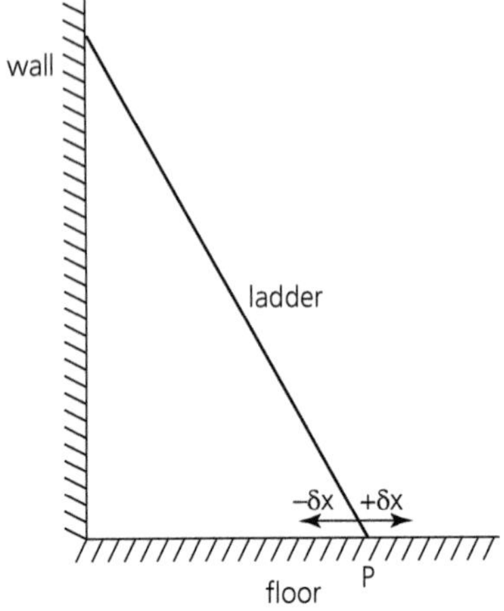

Figure 3.3 Is the ladder stable?

the positive or negative directions (for each i), and this in turn ensures that we can ascertain whether the C-point is at a true stationary point, and not merely at a 'false' extremum existing on a boundary (see the washing line in Figure A3.1, Appendix A3.1).

Now, you may object that, if we are assuming a *stable* ladder, then we have presumed the very answer that we seek. This isn't so, as we are really asking the question: "granted that the ladder is stable, then what are the mathematical conditions that are implied by this stability?" Note that while the displacements are virtual yet they must still satisfy the actual, physical constraint conditions (the ladder-foot moves only within the surface, it doesn't burrow into the floor or jump into the air, and the whole ladder maintains its rigidity).

There's no denying that the idea of virtual displacements is difficult and, at first glance, mysterious, but let's argue the case for mathematical experimentation the other way around. The mathematician can imagine anything he/she wants to (infinite speeds, Dirac delta functions, bishops moving only diagonally, and so on); imagining is not the problem, but why should these physically impossible imaginings tell us anything physically useful, *that* is the question. There are two answers. First, although we have set up this artificial construction of virtual displacements, in fact it is only the point P itself that we're interested in (the position of the C-point in configuration space) and at this point the distinction between dx and δx in effect disappears. Second, there are many precedents where mathematics has provided criteria that have no direct physical manifestation. For example, consider two rods moving in different directions at speeds close to c, the speed of light. Even though the rods move at speeds less than c, the *mathematical* point of intersection of the rods can move faster than c. Quantum mechanics abounds in cases where there are mathematical conditions for which there is no direct physical analogue (although, of course, the mathematics leads on to predictions which *can* be checked up on physically). There is one other telling example, the only other case where the adjective 'virtual' is used in physics. This occurs in geometrical optics, where extensions of actual light rays are mathematically imagined. If these virtual extensions intersect then we have a focused image but it is an image where there is no actual light.

We started this chapter with the broad assertion that "The programme of physics is the mapping of numbers onto physical things…" but now we are finding that the intersection of maths and physics is more murky than this suggests. In fact, the problem of understanding virtual displacements is entirely one of merging the physical and the mathematical worlds: if we were solving a problem in pure mathematics, then all the displacements would be of one kind - virtual - and no problems of understanding would arise.

(3) Imperfect displacements

So far, we have had displacements, dx or δx, and we have been concerned with determining dy or δy, where y is a specific function that has been supplied beforehand. But what if there are displacements but *there is no function y*? For example, we might have a guinea pig nibbling its way through a biscuit: with each tiny bite more and more of the biscuit disappears, but we can't express this as a *functional* relationship between bites and biscuit-shape. All we can say is that each actual, tiny bite, dx, causes an actual, tiny reduction in the biscuit, $\overline{dy}$, called an 'imperfect differential'. We use a bar over the top, $\overline{dy}$, to show that this represents an actual tiny change, but we can't say that it is the 'd-of-y' as we would have said in ordinary calculus. As there is no function, y, then we will not be able to use our usual mathematical weapons (differentiation, etc.). This case looks hopeless - how can it ever be incorporated into such a mathematical subject as the variational mechanics? Lagrange was not defeated, and with amazing ingenuity he devised a method of dealing with these imperfect differentials. But - enough already - we shall not be pursuing this further at this stage.[43]

We said that there were just three ways of taking an infinitesimal step but there is a fourth way - we can make an infinitesimal 'whole path' variation. We'll discuss this in the next section.

3.7.3 The calculus of variations: stationary integral problems - a big step up in complexity

We have been concerned with determining the conditions when a function y is stationary (Section 3.7.2) but, beyond the usual 'regularity' requirements,[44] we have said nothing about the function y itself. In problems of static equilibrium then y is a simple polynomial function (such as $y = ax^2 + bx + c$), but in more general mechanics, where motion is involved, then it turns out that y is generically a lot more complicated - it's then a function-of-a-function-of-a-function. Specifically, y is then an integral, I, and this is a function of the integrand, F, where F is itself a function of f, which is itself a function of the independent variable, x:

$$y \equiv I = \int_a^b F\Big(f(x), \dot{f}(x), x\Big)\, dx \tag{3.8}$$

(Also, as shown in this equation, one of the arguments of F is $\dot{f}$, which is a shorthand notation for the derivative, df/dx - the reason for its inclusion will be explained later on.) The reason why in dynamics we have to do with an integral, and a huge increase in complexity, will be explained in Chapter 6. For now, we concentrate solely on the mathematics. The problem will once again be, as in Section 3.7.2, to determine the virtual variation δy (now also called δI), and to see what conditions are required for it to equal zero. But what is δy when it comes to an integral? It will entail the variation of the 'whole shape' of a function between two end-points, a and b, rather than a displacement made at one 'running' point, x. We first look at some qualitative features of this sort of problem.

[43] If your appetite has been whetted, the method involves a technique known as 'Lagrange multipliers', Section 6.8.

[44] y must be continuous, differentiable, and also finite and single-valued.

Extremum problems of this kind, where the stationarity of an integral is sought, occur in dynamics, and in general mechanics problems. We have met some cases already in Chapter 2: the *brachystochrone* curve; isoperimetric problems (for example, Dido's quest to maximize the area enclosed by a fixed strip of bullhide); Newton's most streamlined solid; and even certain static problems (the equilibrium shape for a suspension bridge - the catenary). In all these cases we must determine the correct overall shape of some continuous function such as to make the integral, I, have a minimum or, more generally, a stationary value, $\delta I = 0$. Once again (as in Section 3.7.2, case (2)), we will use the technique of mathematical experimentation, but this time we will virtually 'vary a function' rather than virtually displace a variable.

What does it mean to 'vary a function'? Looking back at equation (3.8), it seems that there are a number of ways in which the integral could be (virtually) altered: we could change any or all of $F, f, \dot{f}, x$, or the limits a or b. However, in fact, we ban all but one of these changes. First, it is in the nature of these problems that F is never altered, it is prescribed beforehand (in other words, F defines the very problem we are investigating and if we allow it to be changed we are, in effect, investigating a new problem). F might be the formula for an infinitesimal increment of time, ds/v; or the potential energy per unit length of hanging chain; or, in dynamics, as we shall find out in Chapter 6, it is the difference between the kinetic and potential energy functions.[45] Second, we do not allow virtual changes in the independent variable, x. (Of course, x does actually change as we integrate from the limit $x = a$ all the way to $x = b$, but there is never an extra introduced change, δx.) Talking of the limits, a and b, these also are fixed, and so is the value of f at these limits. So, what is left? *The only thing we are allowed to change (to 'vary') is the function, f.* Even $\dot{f}$ is not varied by us, it just changes consequentially as a result of the change in f.

Thus, here is yet another huge step up in complexity: instead of a 'virtual displacement of a variable' we have a 'virtual variation of a function'. This variation is achieved by plucking out of the mathematical imagination an arbitrary function, $\phi(x)$, then scaling it by some 'small' arbitrary constant, ϵ, and adding the result to our starting function, $f(x)$:

[45] Note that it is imperative that F is in *functional* form - if the problem doesn't present itself in this way, then we just have to give up and go home, as there's no way that analytical (i.e. mathematical) mechanics can be applied to this type of problem.

variation of f

$$f^{var}(x) = f(x) + \epsilon\phi(x) \qquad a \leq x \leq b \tag{3.9}$$

The variation can then be made infinitesimal by allowing ϵ to get arbitrarily close to zero. The conditions on $\phi(x)$ are that it must be continuous and differentiable, it must be 'small', and, at the integration limits $x = a$ and $x = b$, it must be zero (in other words, the function is not varied at the ends of the integral). An example of a function, $f(x)$, and some possible variations, $f^{var}(x)$, is given in Figure 3.4. (Terminology and notation: a whole virtual path is known as a 'variation of $f(x)$' and is denoted $f^{var}(x)$; the difference between $f^{var}(x)$ and $f(x)$ at a *given* x is sometimes called the 'variation of $f(x)$ at x' and is denoted $\delta f(x)$, in other words, $\delta f(x) = \epsilon\,\phi(x)$.)

The distinction between an actual differential, df, and a virtual variation, δf, must be borne in mind. While dx and δx are identical mathematically (Section 3.7.2), df and δf are utterly different mathematically. Because f is a function (that is, $f = f(x)$), then df is the usual 'total differential' that we are familiar with from ordinary calculus, and a finite df inescapably implies a finite actual displacement, dx, (see Figure 3.5). On the other hand, the virtual variation, δf, does *not* imply the virtual displacement, δx, but rather δf occurs at an 'instantaneous' value of x. This conforms to the rubric for these stationary integral problems - that a virtual displacement of the independent variable, δx, is not allowed.

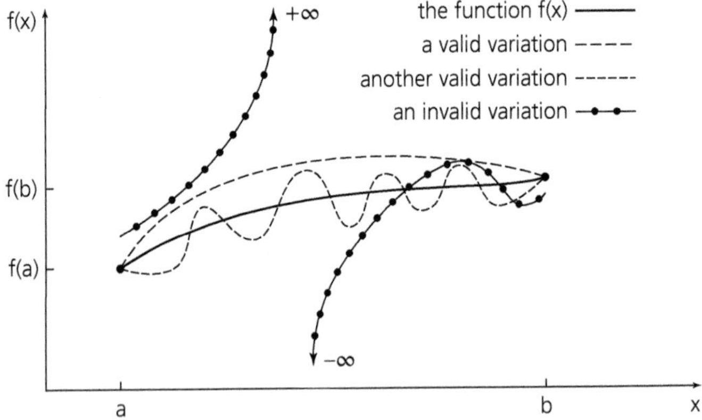

Figure 3.4 The function, $f(x)$, and some variations, $f^{var}(x)$ (one dimension).

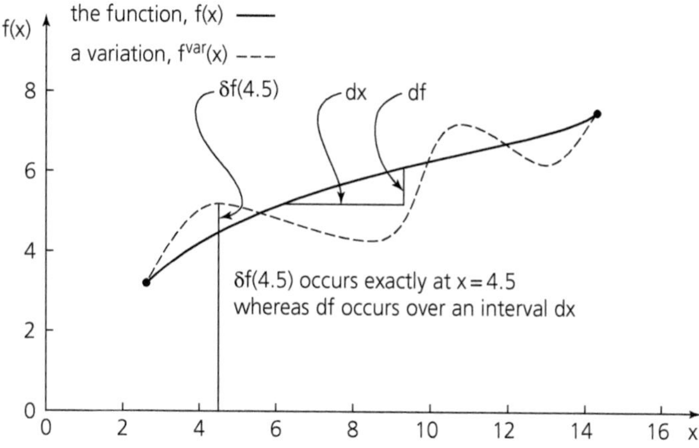

Figure 3.5 The distinction between df and δ (one dimension).

Now that we have established how our integral, I, is to be varied (by infinitesimal virtual variations in f), we can return to our overarching goal - to establish whether the integral is stationary with respect to these variations. In other words, does $\delta I \longrightarrow 0$ as $\delta f(x) \longrightarrow 0$? This is a mathematical problem of considerable difficulty, and it was solved in the eighteenth century first by Euler and then, more rigorously, by Lagrange. In honour of these great mathematicians, the solution is known as the Euler-Lagrange equation, and we shall simply present it rather than show how it was derived:[46]

$$\text{The Euler-Lagrange Equation:} \quad \frac{d}{dx}\left(\frac{\partial F}{\partial \dot{f}}\right) - \frac{\partial F}{\partial f} = 0 \qquad (3.10)$$

$$\text{the solution to} \quad \delta I = \delta\left[\int_a^b F(f(x), \dot{f}(x), x)\, dx\right] = 0$$

with respect to virtual variations, δf

In the case of n dimensions we then have n functions, f_i, and n functions, $\dot{f}_i$, and instead of just one equation, we have a set of n simultaneous equations:

[46] (no more do we show how $d(x^n)/dx = nx^{n-1}$ can be derived).

The Euler-Lagrange Equations

$$\frac{d}{dx}\left(\frac{\partial F}{\partial \dot{f}_i}\right) - \frac{\partial F}{\partial f_i} = 0 \qquad \text{for } i = 1, 2, \ldots n \qquad (3.11)$$

the solution to: $\delta I = \delta\left[\int_a^b F(f_1, f_2, \ldots f_n; \dot{f}_1, \dot{f}_2, \ldots \dot{f}_n; x)\, dx\right] = 0$

with respect to virtual variations, $\delta f_1, \delta f_2, \ldots, \delta f_n$

There is still one loose end to tie up. At the beginning of this section we mentioned that one of the arguments of F is $\dot{f}$ - why must this be so? (It would seem at first sight that as the function f is determined by solution of the Euler-Lagrange equation then $\dot{f}$ is also so-determined, and therefore $\dot{f}$ need not be specified separately as an argument?) Our frequent remark that the functions f and δf (and, by implication, F and δF) must be continuous has often been relegated to brackets and footnotes, but now we can see that this condition is crucial: there will be an infinite number of ways in which a path could be stationary and discontinous, but only *one* way in which it could be stationary, satisfy the end-conditions, and still be continuous.[47] In effect, the requirement of continuity is like lots of extra 'boundary' conditions all along the length of the path. The boundary conditions at the very ends of the varied path are provided by the upper and lower limits of the definite integral, however the 'internal boundary conditions' are provided by the function $\dot{f}$ - the value of $\dot{f}$ at any given x ensures the continuity of f at that x. (This is mentioned again in Section 6.4, Chapter 6.)

We have been laying the mathematical groundwork - all will become clearer when the physical motivation for this mathematics is explained, and some worked examples are given (Chapter 6 and Appendix A6.1).

[47] (This continuity condition is an especially intuitive requirement when we are dealing with a vibrating string, a hanging chain, and so on.)

4

The Principle of Virtual Work

4.1 Introduction

Classical Mechanics deals with 'statics' (nothing is moving) and 'dynamics' (some parts are moving relative to other parts). We consider, in this chapter, the first of these.

If nothing is moving then the system is said to be in a state of static equilibrium. We have two methods of dealing with such static states, the method of Newtonian Mechanics, and the method known as the Principle of Virtual Work. Physicists have tended to use the first method, whereas engineers have used both methods, as appropriate. Thus the engineers are ahead of the physicists, showing the way for the last 150 years and more (the Principle of Virtual Work was first proposed by Johann Bernoulli in 1715 (see Chapter 2)).

There is a difficulty in comprehending these static states: how shall we know whether there are strong forces keeping a certain structure in balance - a Roman arch, a house of cards, taut cords meeting at a point - or whether the forces are weak, or even absent, and the various parts just happen to be adjacent to each other in space? Similarly, what is the evidence for the large amount of energy locked away in the still body of water in the reservoir of a hydroelectric power station, a column with a heavy weight on top, or a fully charged battery? Our instinctive response is to gently prod, shake, or tweak the system, trying to disturb it very slightly away from equilibrium. In this way, the forces and energies can reveal themselves. We shall find that this nudge to the system is pursued in a special way in the Principle of Virtual Work.

4.2 System of non-interacting particles

We start by considering a very simple scenario - a system made up just of forces and non-interacting particles. Particles are point masses,

The Lazy Universe. Jennifer Coopersmith, Oxford University Press (2017).
 DOI 10.1093/acprof:oso/9780198743040.001.0001

$m_1, m_2, \ldots$, that is, they have no extension and no internal structure (but they may have extra properties such as electric charge). Forces are applied externally, such as gravity, muscle strength, and so on. We seek to determine the magnitude of the forces necessary to achieve equilibrium.

Method 1, Newtonian Mechanics

Consider that we have particles $i = 1$ to N, with position vectors, $\mathbf{r}_1, \mathbf{r}_2, \ldots \mathbf{r}_N$, acted upon by forces, $\mathbf{F}_1, \mathbf{F}_2, \ldots \mathbf{F}_N$ respectively (the $\mathbf{r}_i$ and $\mathbf{F}_i$ are referred to the usual Cartesian axes (x, y, z)). To say that the system is in equilibrium is to say that the particles are not moving. But if nothing is moving then it is also true that nothing is *accelerating*; so, by Newton's Second Law of Motion, there are no forces. Therefore we must have:

Method 1, static equilibrium of a system of N non-interacting particles in Newtonian Mechanics

$$\mathbf{F}_1 = 0, \quad \mathbf{F}_2 = 0, \quad \ldots, \quad \mathbf{F}_N = 0 \tag{4.1}$$

This is a condition involving forces but what of the particles? And are there the same number of forces as particles? For answer, we note that: if there happens to be a force where there is no particle (the force acts in empty space somewhere) then this force is not relevant to the problem and can be ignored; if there happens to be a particle where there is no force then we can just as well say that there *is* a force but its magnitude is zero; and if there happen to be two or more forces acting at a given particle then the forces can be added to yield just one resultant force. On all counts (4.1) still holds and so, in all generality, the equilibrium condition for N non-interacting particles has N forces, one force acting at each particle, and each force is zero. It's as simple as that.

Method 2, The Principle of Virtual Work

The Principle of Virtual Work (P of VW) tackles the problem of static equilibrium in an entirely different way. The Principle first contrives a scenario in which each and every particle, i, simultaneously does an infinitesimal amount of virtual work, $\delta\omega_i$. Leaving aside, for the moment, how this is contrived, and what 'virtual' means, note that 'work' is still defined, from Classical Mechanics, as the dot product of force and displacement. $\delta\omega_i$ is therefore a scalar quantity, but one that can be positive or negative. ('Positive' when, along any given axis, the force and displacement vectors point the same way; 'negative' when, along any given axis, the force and displacement vectors point the opposite way.) The P of VW then asserts that the system will be in static equilibrium if, after summing these simultaneous contributions to virtual work, one contribution from each particle, the total virtual work is zero:

Method 2, static equilibrium of a system of N non-interacting particles in the Principle of Virtual Work

$$\delta\omega_1 + \delta\omega_2 + \ldots = \sum_{i=1}^{N} \delta\omega_i = \delta\omega^{total} = 0 \tag{4.2}$$

In striking contrast to (4.1), forces are nowhere to be seen.

We now return to an explanation of what is virtual work. 'Work' is defined in the usual way as the 'work done by a particle acted upon by a force while the particle is displaced through a certain distance in a certain direction.' But Method 2 immediately presents us with a problem: in order to have finite contributions to work we need finite displacements but, at equilibrium, nothing is moving and so the displacements are all zero. Therefore, to rescue Method 2, we do something which at first sight appears like an act of intellectual prestidigitation - we invoke a small, *hypothetical* displacement, $\delta\mathbf{r}_i$, one for each particle, i. This is the 'nudge' that was referred to at the end of the introduction. The symbol, δ, pronounced 'variation', reminds us that the displacements (and the consequent 'work done') are imagined, virtual, and form part of a purely mathematical experiment.

At first encounter this seems perplexing - why do we not actually, *physically*, tweak the system (say, use our finger to move a particle)? However, strange as it may seem, we shall find (Sections 4.5 and 4.9) that some virtual displacements cannot occur physically, and some physical displacements cannot occur virtually. We must simply accept the need for virtual displacements - imagined displacements that happen in an imagined mathematical landscape or 'space'. This 'space' is the 'configuration space' that was described in Chapter 3.

Conditions (4.1) and (4.2) appear utterly different, yet we may suspect that Methods 1 and 2 really boil down to the same thing. After all, work is the scalar product of force and displacement, and so if all the forces are zero then surely the work must also be zero, end of story? However, this innocent line of reasoning turns out to be flawed. Methods 1 and 2 really are quite different, and for at least three reasons. For one thing, they concern totally different entities (Method 1 has to do with forces, Method 2 has to do with energy, specifically work), and for another, Method 1 involves lots of individual statements ($\mathbf{F}_1 = 0$, $\mathbf{F}_2 = 0, \ldots$) while Method 2 involves one grand statement which is a sum. Finally, while Method 1 involves equating things to zero, Method 2 involves a mathematical procedure which goes beyond mere summing to zero and uses the variational calculus to carry out a 'test of flatness' or 'test of stationarity' (Chapter 3). We shall talk more about this later.

We would nevertheless like to bring out the resemblances between the two methods, and so we re-express (4.2) in terms of forces and displacements:

Static equilibrium in the Principle of Virtual Work, re-expressed

$$\mathbf{F}_1 \cdot \delta\mathbf{r}_1 + \mathbf{F}_2 \cdot \delta\mathbf{r}_2 + \ldots = \sum_{i=1}^{N} \mathbf{F}_i \cdot \delta\mathbf{r}_i = \delta\omega^{total} = 0 \qquad (4.3)$$

The forces, $\mathbf{F}_1$, $\mathbf{F}_2$, …, are external and have been supplied in the specification of the given problem. Despite these forces, the particles in this example are 'free' (they don't interact with each other, and are not constrained) and so each is free to be displaced in *any* direction. We can therefore choose the virtual displacement for each particle independently, and to be in any direction. So much for the direction but what of the magnitude of the displacement? We must understand that (4.3) is more than an equation, it is a 'test of stationarity' and therefore represents a limiting process. The sum must be zero *in the limit* as the magnitude of each displacement is reduced. The starting magnitude must be 'small' but it doesn't matter what exactly it is (in the same way as, in differential calculus, we draw diminishing triangles on a curve in order to find the slope, but it doesn't matter what size triangle we start with). So, for these independent particles, we can choose both the direction and (small) starting magnitude of the virtual displacements at random. Condition (4.3) is therefore a condensed version of the following infinite set of equations, each one corresponding to a different choice or set of virtual displacements:

$$\sum_{i=1}^{N} \mathbf{F}_i \cdot \delta \mathbf{r}_i^{choice1} = 0 \tag{4.4}$$

$$\sum_{i=1}^{N} \mathbf{F}_i \cdot \delta \mathbf{r}_i^{choice2} = 0 \tag{4.5}$$

$$\sum_{i=1}^{N} \mathbf{F}_i \cdot \delta \mathbf{r}_i^{choice3} = 0 \tag{4.6}$$

$$\text{and so on} \quad \vdots$$

The only way that this infinity of equations can be simultaneously satisfied is if all the forces are zero, $\mathbf{F}_1 = 0$, $\mathbf{F}_2 = 0$, ..., $\mathbf{F}_N = 0$. We have returned to exactly the same result as the one derived from Method 1.

We see that Method 2, the Principle of Virtual Work, involves a much more complicated procedure in order to end up at exactly the same results as Method 1. Before explaining why this is justified, let's check whether the Newtonian predictions are again reproduced by the P of VW when more complex equilibrium states are considered. Suppose we now allow for the particles to interact - to be bound together into a rigid body, for example.

4.3 Statics of a rigid body

What is the difference between a rigid body and a collection of particles? Lanczos poses the question most graphically: what is the difference between a sandstone rock and a pile of sand just happening to have exactly the same overall shape as the rock? He answers that we can try picking up 'the body' and moving it from one place to another (giving it a translation or a rotation). The rock will move as a whole, as a rigid body, whereas the sand-pile will disintegrate. This translation or rotation is a kind of physical nudge, but the body will only remain intact if the nudge is sufficiently gentle. Likewise, in applying the test of stationarity in the P of VW, we give the body a gentle *virtual* nudge. We now describe how this is achieved.

We can make the adjective 'gentle' more precise. The rock is rigid because there are internal forces (also called constraint, or reaction forces) between the particles, binding them together into a body. The virtual displacements will be sufficiently 'gentle' if they never go against these forces of reaction, and this will be the case if the body is always displaced (virtually) *as a whole*, and is never virtually squeezed, twisted, or stretched. This is guaranteed so long as we give every particle exactly the same virtual translation:

$$\delta \mathbf{r}_i = \delta \mathbf{r} \quad \text{for all particles} \quad i = 1 \text{ to } N \tag{4.7}$$

The resultant total virtual work due to these virtual translations is then:

$$\sum_{i=1}^{N} \mathbf{F}_i \cdot \delta \mathbf{r} = \delta \mathbf{r} \cdot \sum_{i=1}^{N} \mathbf{F}_i = \delta \omega^{total} = 0 \tag{4.8}$$

The forces, $\mathbf{F}_i$, are not the internal ones, but are the external forces that we seek to determine, and that ensure the equilibrium state. As $\delta \mathbf{r}$ is non-zero, (4.8) can only be satisfied if:

$$\sum_{i=1}^{N} \mathbf{F}_i = \mathbf{F}^{resultant} = 0 \tag{4.9}$$

But this is just the standard result from Newtonian Mechanics: *for equilibrium of a rigid body, the resultant external force must be zero.*

A new development arises now that we are dealing with bodies rather than particles. A rigid body, having extension and having a shape, can be rotated.[1] This possibility brings in a new kind of virtual displacement - not just translations but also rotations. Again, the reaction forces must not come into (virtual) play, and this is assured by giving every particle (that is, every particle's position vector, $\mathbf{r}_i$) exactly the same virtual rotation about some common axis of rotation:

$$\delta \theta_i = \delta \theta \quad \text{for all particles } i = 1 \text{ to } N \tag{4.10}$$

However, unlike (4.7), this universal rotation does still lead to different displacements, as these depend on the individual particle position vectors, as follows:

$$\delta \mathbf{r}_i = \delta \theta \, \mathbf{U} \times \mathbf{r}_i \tag{4.11}$$

[1] As a particle has no size then there is no evidence that it has been rotated. This is equivalent to saying that it cannot be rotated.

where **U** is a vector of unit length along the given axis of rotation. The virtual work of particle, i, due to the external force, $\mathbf{F}_i$, acting at i, becomes:

$$\delta\omega_i = \mathbf{F}_i\cdot(\delta\theta\,\mathbf{U}\times\mathbf{r}_i) = \delta\theta\,\mathbf{U}\cdot(\mathbf{r}_i\times\mathbf{F}_i) = \delta\theta\,\mathbf{U}\cdot\mathbf{M}_i \qquad (4.12)^2$$

where $\mathbf{M}_i = (\mathbf{r}_i\times\mathbf{F}_i)$ is known as the 'moment of the force' about the given axis of rotation. Therefore the total virtual work due to all particles is:

$$\delta\omega^{total} = \sum_{i=1}^{N}\delta\theta\,\mathbf{U}\cdot\mathbf{M}_i = \delta\theta\,\mathbf{U}\cdot\sum_{i=1}^{N}\mathbf{M}_i \qquad (4.13)$$

As $\delta\theta$ **U** is non-zero, the total virtual work can only be zero if:

$$\sum_{i=1}^{N}\mathbf{M}_i = \mathbf{M}^{resultant} = 0 \qquad (4.14)$$

which is the same as saying that **U** must be perpendicular to $\mathbf{M}^{resultant}$. Again, this is the same as the result we expect from Newtonian Mechanics: *for equilibrium of a rigid body, the resultant moment is zero.*

4.4 A comparison between Newtonian Mechanics and the Principle of Virtual Work

The very simple scenarios in Sections 4.2 and 4.3 have been selected to demonstrate the complete equivalence of Newtonian Mechanics and the Principle of Virtual Work in these cases. It appears that the P of VW has served only to replicate Newtonian Mechanics, and at the cost of introducing obscure imaginary motions in a virtual space - but we should remember that the P of VW has arrived at these results, conditions (4.9) and (4.14), afresh and completely independently of Newtonian Mechanics. In order to bring out the new outlook and content of the P of VW we try a different tack: instead of introducing forces into (4.2) we introduce displacements into (4.1). Also, we consider a

[2] We have used the result from vector algebra that $\mathbf{a}\cdot(\mathbf{b}\times\mathbf{c}) = \mathbf{b}\cdot(\mathbf{c}\times\mathbf{a})$.

more general system in which particles can interact but are not necessarily bound into a rigid body. Each particle may be subject to an internal force, an external force, or both, in other words, each particle is acted upon by a net force, $\mathbf{F}_i^{net}$, which is the sum of the applied and constraint forces at i. (We shall in what follows use the terms 'applied' for 'external', and 'constraint' for 'internal'.) The Newtonian condition for equilibrium (Condition (4.1)) can then be written:

$$\mathbf{F}_1^{net} = 0, \quad \mathbf{F}_2^{net} = 0, \quad \ldots, \quad \mathbf{F}_N^{net} = 0 \tag{4.15}$$

As the net force at each particle is zero, we are able to take its scalar product with any vector we like and still obtain zero. Exploiting this fact, we choose to take the scalar product of $\mathbf{F}_i^{net}$ with whatever happens to be the virtual displacement at i. For this special choice-set, we arrive at:

$$(\mathbf{F}_1^{net} \cdot \delta\mathbf{r}_1) = 0, \quad (\mathbf{F}_2^{net} \cdot \delta\mathbf{r}_2) = 0, \quad \ldots \quad (\mathbf{F}_N^{net} \cdot \delta\mathbf{r}_N) = 0 \tag{4.16}$$

Adding lots of zeros together should still lead to zero, so we obtain:

$$(\mathbf{F}_1^{net} \cdot \delta\mathbf{r}_1) + (\mathbf{F}_2^{net} \cdot \delta\mathbf{r}_2) + \ldots + (\mathbf{F}_N^{net} \cdot \delta\mathbf{r}_N) = 0 \tag{4.17}$$

Remembering that $\mathbf{F}_i^{net}$ is made up from an external applied force, $\mathbf{F}_i^{appl}$, and an internal constraint force, $\mathbf{F}_i^{cons}$, equation (4.17) becomes:

$$(\mathbf{F}_1^{appl} + \mathbf{F}_1^{cons}) \cdot \delta\mathbf{r}_1 + \ldots + (\mathbf{F}_N^{appl} + \mathbf{F}_N^{cons}) \cdot \delta\mathbf{r}_N = 0 \tag{4.18}$$

But now, instead of considering everything on a particle-by-particle basis, we rather lump together all the $\mathbf{F}_i^{appl}$s, and lump together all the $\mathbf{F}_i^{cons}$s. In this way we arrive at:

$$\sum_{i=1}^{N} (\mathbf{F}_i^{appl} \cdot \delta\mathbf{r}_i) + \sum_{i=1}^{N} (\mathbf{F}_i^{cons} \cdot \delta\mathbf{r}_i) = 0 \tag{4.19}$$

At this stage we introduce a brand new requirement, that the second term in (4.19), that is, the total virtual work of all the constraint forces, is zero:

$$\sum_{i=1}^{N} (\mathbf{F}_i^{cons} \cdot \delta \mathbf{r}_i) = 0 \tag{4.20}$$

This is beautifully consistent: the total 'constraint' (for example, the 'constraint' of being a rigid body) has no internal movements within it, and so it also is a system in equilibrium, a mini-system to which the P of VW should apply.

As we still have equation (4.19) to be satisfied, along with (4.20), then it must be that the total virtual work of all the applied forces is also zero. This, finally, is the P of VW as it is usually defined:

The Principle of Virtual Work, standard definition

$$\sum_{i=1}^{N} (\mathbf{F}_i^{appl} \cdot \delta \mathbf{r}_i) = 0 \tag{4.21}$$

Let us remind ourselves of (4.15), the Newtonian condition we started from:

Equilibrium Condition, Newtonian Mechanics

$$\mathbf{F}_1^{net} = 0, \quad \mathbf{F}_2^{net} = 0, \quad \ldots, \quad \mathbf{F}_N^{net} = 0 \tag{4.22}$$

Comparing these two conditions, (4.21) and (4.22), we find that (4.21) is making an utterly different claim to (4.22). In the P of VW there is just one zero condition, that condition is a sum, and constraint forces are conspicuous by their absence. Also, curiously, there is no longer a necessity for Newton's Third Law of Motion (see Appendix A1.1) to be upheld (at the level of particle-particle interactions). This is because the sum in (4.21) means that the applied forces act as *one system*, there is no extra requirement for certain pairs of forces to cancel each other out. With the foreknowledge that physics contains many examples where Newton's Third Law may be broken,[3] there is every reason why the P of VW, rather than Newtonian Mechanics, will apply in these domains.

[3] For example, the interaction between two electrons, one moving along a line at right angles to the motion of the other. See Feynman, Richard, *The Feynman Lectures on Physics*, Addison-Wesley, 1963, Volume II.

4.5 Virtual Displacements

We have talked much of forces and little about displacements but both occur in (4.21). The virtual displacements, as we have discussed above (Section 4.2, and 3.7.2) are displacements in an abstract space (configuration space), a space with axes that correspond to the degrees of freedom of our given problem. The 'test of stationarity' occurs at one point - the C-point - of this space. Although the virtual displacements are imagined this doesn't mean we are free to imagine anything we like. We must follow some strict guidelines, as follows:

How many virtual displacements are there?

The forces are given in the description of the problem, but the virtual displacements are chosen by us.[4] They are chosen to occur wherever there is an applied force acting on a particle. If, by chance, we happen to do a virtual displacement where there isn't a force, it won't matter (it's virtual work will be zero), but if there is an applied force acting on a particle somewhere and we don't carry out a virtual displacement at this position then the 'test of stationarity' (the Principle of Virtual Work) will fail, it won't be 'complete'.

When do the displacements occur?

The 'test of stationarity' is carried out at a test *point*, in other words, in an instant, that is to say, at *one* time. Therefore all the virtual displacements must happen simultaneously.

Where do the displacements occur?

Each virtual displacement, $\delta\mathbf{r}_i$, must be located at the end of the respective vector, $\mathbf{r}_i$.[5]

What direction does the displacement have?

The direction of $\delta\mathbf{r}_i$ is arbitrary except that *it must be 'in harmony' with any constraints*, which is the same thing as saying that it must be perpendicular to

[4] In some advanced versions of the method of VW, the applied forces can be given a virtual variation instead of the displacements; we won't be considering these versions.

[5] The notation makes $\delta\mathbf{r}_i$ look similar to $\mathbf{r}_i$ whereas in fact they are quite different: $\delta\mathbf{r}_i$ is a displacement vector while $\mathbf{r}_i$ is a position vector.

the reaction forces at i (see Section 4.8). Only in this way we can be sure that no virtual work will arise from these constraint or reaction forces. Also, for any allowed direction, the virtual displacement could be $+\delta\mathbf{r}_i$ or $-\delta\mathbf{r}_i$ (the displacements are said to be 'reversible') and this ensures that the test point can be approached from the negative or positive side of any allowed direction. (See Appendix A3.1.)

What magnitude do the displacements have?

As mentioned before, we must appreciate that condition (4.21) is really a shorthand for a mathematical procedure ('taking the limit'). For independent particles, each virtual displacement can have any starting magnitude, except that this magnitude must always be 'small' - as we can only examine a *local* region surrounding the test point. The 'test of stationarity' is then applied by simultaneously letting all the imagined displacements tend to zero-magnitude as the test point is approached. If there are functional relations between the position vectors (relations given in advance in the description of the problem) then the virtual displacements will also have to satisfy these relations, as determined by the usual rules of differential calculus.[6]

How long do the displacements take?

As mentioned above, the 'test of stationarity' means that the displacements must happen simultaneously. They must also happen instantaneously, that is, take no time to occur.[7] As they are imagined rather than actual, this doesn't pose a problem.

In summary, 'virtuality' means that the $\delta\mathbf{r}_i$ are 'small', happen simultaneously, and do not cause a force, result from a force, or take any time to occur. Also, they can only be in directions that are 'in harmony' with any constraints and other 'kinematical conditions' that may pertain. You may wonder why the displacements can depart from physicality as regards their duration, yet must bow to it as regards constraints and kinematical conditions? The answer is that, out of an infinity of possible configuration spaces, we wish to investigate this one and not that one (that is, we wish to investigate the one that does correspond to our given system).

[6] So, if, say, the given problem requires that $r_1 = (r_2)^3$, then we must have $\delta r_1 = 3(r_2)^2\delta r_2$. Also, if, say, $\mathbf{r}_1 = \mathbf{r}_2 + \mathbf{r}_3$, then $\delta\mathbf{r}_1 = \delta\mathbf{r}_2 + \delta\mathbf{r}_3$.

[7] This is an example of virtual displacements which have no actual, physical, counterpart - mentioned in Method 2, Section 4.2.

4.6 The Principle of Virtual Work: Feynman's pivoting bar

It will be easier to understand these abstract ideas if we use the P of VW to tackle a specific problem. We consider an example adapted from Feynman's Lectures on Physics.[8] A bar, 8 metres long, is supported on a fulcrum at one end (Figure 4.1). In the middle of the bar is a mass of 60 kg, and at a distance of two metres from the support is a mass of 100 kg. What value must the hanging mass, M, have in order that the bar is balanced (is in equilibrium)?

The three masses can be considered as 'particles', and the bar introduces a constraint or kinematical condition between these particles (assume that the mass of the bar itself is negligible, that it remains rigid while it pivots, and that there is no friction at the pivot or in the sliding of the pulley-cord). We imagine that M falls any arbitrary 'small' distance, say, 4 cm. The centre mass then rises 2 cm, and the other mass rises 1 cm. So we have virtual displacements −4 cm, +2 cm, and +1 cm, and weights (applied forces) of Mg N, 60 g N, and 100 g N, respectively (where 'g' is the magnitude of the acceleration due to gravity). Thus, the P of VW, (4.21), leads to:

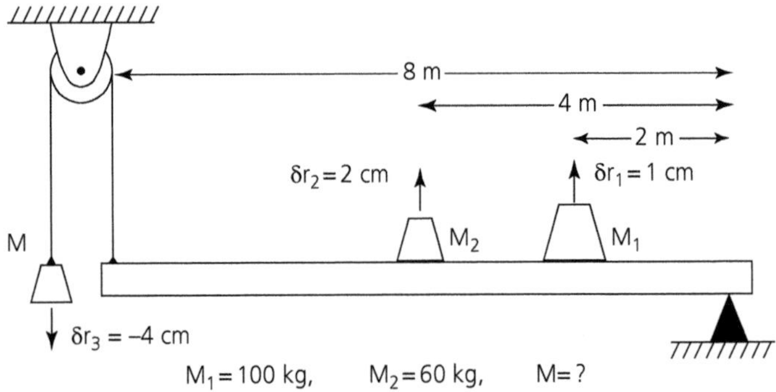

Figure 4.1 Weighted bar supported at one end (adapted from Fig. 4–6, Vol I of The Feynman Lectures on Physics, Fifth printing, 1970, courtesy of California Institute of Technology, © in 1963.)

[8] R. P, Feynman, Chapter 4, Vol I, *The Feynman Lectures on Physics*, 1963. Feynman was not only one of the outstanding physicists of the twentieth century, but was also one of the best explainers of physics.

$$(Mg \times -4) + (60\,g \times 2) + (100\,g \times 1) = 0 \tag{4.23}$$

and so we find that $M = 55$ kg.

As Feynman explains, the beauty of the P of VW is that we can try out imaginary displacements even though in practice the bar may be immoveable.

By the way, the first virtual displacement was chosen to be 4 cm; this is not infinitesimal but is it 'small enough'? Answer: yes, as on this scale all the virtual displacements approximate to straight-line segments.[9] So 'sufficiently small' depends on details of the actual physical scenario, and on how precise the measurements are.

4.7 Coordinates - an increase in generality

First we considered free particles, then particles constrained into a rigid body, then particles constrained any which way. Now we take yet another step towards increased generality, but this is not a small step, it is enormous, and will allow us to see the whole of physics in a new way. We will model the physical world employing not necessarily 'particles' but 'things'; not 'displacements' but 'motions'; not 'applied forces' but 'applied generalized forces'. It is the far reach of the Principle of Virtual Work that gives us this licence; and as the crucial elements are now infinitesimal chunks of *energy* (condition (4.2)), so the crucial sub-elements are any quantities that combine together to yield *energy*.

It's time to come clean; from the start of this chapter we have lulled you into a false sense of universality, and taken it for granted that the physical system is always reducible to forces, particles, and the position vectors of these particles. Also, we casually stated that the position vectors and forces were 'referred to the usual Cartesian axes, x, y, and z' (Section 4.2). But all the above assumptions are inessential and prejudicial. The Cartesian coordinates are 'rectangular' (Chapter 3) but as soon as the physical system involves a rotation, a bend, a twist, spinning, orbiting, pivoting, spherical bodies, bubbles - anything *curvy* - then rectangular coordinates are a poor choice, perhaps even an impossible one. Although vectors offer such wonderful insights, and greatly facilitate the ease of solving problems such as 'the equilibrium of taut cords meeting at a point', 'a swimmer swims diagonally across a uniformly flowing

[9] In actual fact, the displacements are curved, being small arcs of circles traced out as the bar pivots.

river', and so on, nevertheless they lead to fiendish complications whenever curvy features are present.

Why didn't we notice this before? It is partly because, in our physics and engineering textbooks, the problems have been very carefully selected, one might even say cherry-picked. Also, some standard physics equations, equations of vector algebra such as $\mathbf{F}_1 + \mathbf{F}_2 = \mathbf{F}^{resultant}$, and $\mathbf{F} = m\mathbf{a}$, are so simple, so seemingly elemental, that we can hardly credit that they are, actually, rather specialized. For example, $\mathbf{F} = m\mathbf{a}$ applies only to *linear* momentum.[10] In fact, components like 'particle', and 'force', and equations like '$\mathbf{F}_1 + \mathbf{F}_2 = \mathbf{F}^{resultant}$', are elemental, but this doesn't prove that they are the only elements or the right elements in a given physics problem.

We therefore need to stop presuming Cartesian coordinates (x, y, z), and, depending on the given scenario, we could use, say, the spherical polar coordinates, (r, θ, ϕ).[11] This is the perfect coordinate system for certain problems (for example, the spherical pendulum) but we have in mind a more radical change than a mere switch from one coordinate system to another. For instance, there is no need to think only in terms of positioning *particles*, we could just as well map the motions of other system-components, including whole entities such as 'lever arm', 'cricket bat', 'juggling club', 'a spring', 'a sliding block', 'a planet', 'a piston rod', and so on. Finally, we can free ourselves from thinking of 'motions' as just translations or rotations, and consider also changes in capacitance, surface tension, magnetic field, phase of a wave, strain in a beam, pressure within a fluid, and so on. In fact, any variable that can be quantified, is expressible as a function, and characterizes the physical system, can serve as a coordinate of that system. We have arrived at the generalized coordinates, q_i, of Chapter 3.

The forces must be generalized as well. As the displacements are no longer necessarily lengths (quantities with units such as cm, or inches) then the new 'generalized forces' are no longer necessarily Newtonian forces (measured in Newtons). Thus these generalized forces, symbolized Q_i, may have units of Newtons, Newton metres, Newton/(metre)^2, Volts, Joules, and so on. The q_i corresponding to these Q_i could then have units of, respectively: metres, pure number, (metre)^3, Coulombs,

[10] The analogue to this equation for non-linear momentum is: torque, $\mathbf{N} = m\frac{d}{dt}(\mathbf{r} \times \mathbf{v})$, where m, $\mathbf{r}$, and $\mathbf{v}$, are the mass, position, and velocity, respectively.

[11] (The r here has nothing to do with the magnitude of the rectangular vector, $\mathbf{r}_i$.)

angle in degrees, and so on. The important thing is that, as we are performing a test of stationary *work*, so each duo must form a product, $Q_i q_i$, that yields the scalar quantity having the dimensions of *energy*.

How shall we know that we have enough generalized coordinates and the right ones for the given system? This is where the wisdom and artistry comes into mechanics - there is no prescription, no hard and fast rule for choosing the q_i. There isn't even necessarily a unique way of choosing them. The best that can be said is that when some different ways of describing the system have been tried then that choice (or choices) with the smallest number of coordinates is rather special as this smallest number is equal to the 'number of degrees of freedom' of the given system. The 'degrees of freedom' have the special property that they represent *independent* characteristics ('motions') of the system. If they were not independent then there would be some equation linking them (for example, a constraint equation) and then these $\{q_i\}$ would no longer be the smallest set.

It seems that the degrees of freedom (and the number of them) is an ineffable thing, a characteristic of the system rather than of any mere description of it. However, we don't need to feel daunted by the burden of choosing a set of q_i - if we don't notice an effect or characteristic then it probably isn't noticeable (on the scales and with the precision that we are considering)- see Section 3.3. Also, if we have too many q_i, then that also isn't a problem; we'll find, as mentioned above, that the q_i will be linked by some constraint equations or kinematic conditions. An example is Feynman's bar (Figure 4.1, Section 4.6). There are three masses ('particles') and correspondingly three virtual displacements, and so we might be tempted to think that this is a problem with three degrees of freedom. On second thoughts, we realize that these displacements are linked together by a constraint condition (the 'rigidity of the bar') which determines that when mass M moves down by 4 cm the middle mass must move up through 2 cm and the end mass must move up through 1 cm. In other words, the bar is capable of only *one* independent motion - it can rotate about the pivot in the vertical plane. So there is just one truly independent virtual displacement, $\delta\theta$, that is, there is just one degree of freedom. The P of VW condition can this time be written as $\delta\theta[(Mg \times -4) + (60g \times 2) + (100g \times 1)] = 0$, where now the whole expression, '$[(Mg \times -4) + (60g \times 2) + (100g \times 1)]$', is the generalized force, Q. The $\delta\theta$ is finite (there is no purpose in imagining a zero displacement) and so the total virtual work will only equal zero

Table 4.1 Systems, degrees of freedom, and possible q_i

---

one dof: a piston moving up and down; a bead moving along a wire, a bar pivoted at one end, a bubble expanding or contracting, capacitor plates moving closer or further apart,

two dofs: a particle moving on a surface (x,y) or (r,θ); Meriam and Kraige's 'black box' and pushrods A and B (see Section 4.10, problem (i)),

three dofs: a particle moving in space (x,y,z) or (r,θ,ϕ); a rigid body rotating about a fixed point (ω,θ,ϕ),
four dofs: a double star rotating in a plane (x_1,y_1,x_2,y_2) or $(r_1,\theta_1,r_2,\theta_2)$,

five dofs: two particles at a constant distance $(x_1,y_1,z_1,x_2,y_2,z_2)$ + constraint equation, or $(x_{cm},y_{cm},z_{cm},\theta,\phi)$ where 'cm' are the centre-of-mass coordinates,

six dofs: a rigid body moving freely in space (x,y,z,r,θ,ϕ); three non-interacting particles moving on a surface $(x_1,y_1,x_2,y_2,x_3,y_3)$; two free particles in space $(x_1,y_1,z_1,x_2,y_2,z_2)$ or $(r_1,\theta_1,\phi_1,r_2,\theta_2,\phi_2)$.

---

if $Q = 0$, and we return to our earlier equation, (4.23). We have lost the elemental simplicity of **F** and now have Q, a complicated scalar quantity that must be formulated afresh for each new problem. However, we have gained the ability to solve every mechanical problem, and with whatever choice of (q_i,Q_i), and with just one universal principle - the Principle of Virtual Work.

Some examples of mechanical systems, and the generalized coordinates and number of degrees of freedom ('dof'), are given in Table 4.1.

4.8 Constraints and kinematical conditions

As well as choosing the generalized coordinates and ascertaining what are the generalized forces, we must see if the problem involves constraints or other kinematical conditions. The condition, 'rigid body', is an example of a constraint (each internal particle is constrained to stay at a certain fixed position relative to the other internal particles); and 'ideal joint' is another example (components are constrained to maintain certain positions and orientations relative to each other). The constraint may enter into the problem as a constraint equation, or it may be implicit in the mathematical form of the generalized force (e.g. Feynman's pivoting bar, Sections 4.6 and 4.7).

Constraints sometimes manifest themselves as kinematic conditions, such as: a block slides along a table-top (the block is constrained to remain within a given surface); a bob swings at the end of an inextensible cord (it is constrained to keep a certain distance from the ceiling-attachment-point); and so on. It sounds like geometry but a constraint or kinematic condition is never a matter of pure geometry - in the last resort, it is always made up from forces. Moreover, in the last resort these internal or constraint forces are always forces of reaction. For example, why doesn't the sliding block burrow down into the surface of the table? It is because the forces between the atoms in the table oppose this burrowing motion. Why does the rigid body maintain its fixed shape? Answer: because the atoms which make up the body resist, by *force*, being pushed closer or stretched further apart.

This seems like a step backwards - earlier we dispensed with forces (except in simple 'rectangular' vectorial problems) but now we are reintroducing them. The explanation is as follows. There are in mechanics two very different kinds of forces and they are dispensed with in two very different kinds of ways. In the first place we have the external applied forces, and these disappear by becoming the generalized forces; in the second place we have the internal constraint forces, and these disappear (don't come into play) by choosing our virtual displacements very carefully - they must never 'go against' the constraints (see also Section 4.9). The applied forces (for example, gravity) are mathematical functions, specified in advance, or determined when the problem has been solved. The constraint/internal/reaction forces are microscopic in origin, usually of strength unknown, are not given in mathematical form, and can be completely ignored *if* the problem is formulated in the right way: the q_i must be well-chosen, and the δq_i must be 'in harmony' with the constraints. For example, a pendulum bob can be virtually displaced such as to make the pendulum swing ('harmonious'), but not such as to make the pendulum cord stretch (not 'harmonious'). Also, a lever arm can virtually rotate gently about the fulcrum ('harmonious'), but it must not flex or get knocked off its perch (not 'harmonious').

If we push a body, what makes it move? For example, if we grab a loaf of bread at one end, what makes it move at the other end? If the bread was sliced, would the outcome be different? Again, it is the infinity of internal forces that come to our aid: they 'transmit' our applied force from the atoms at one end to the adjacent atoms, and so on, all the way along the length of the bread. The loaf will move as one body, and maintain its shape, provided that the forces are transmitted infinitely fast, and provided that all the internal, inter-particle forces depend only on relative distance (the forces are 'central'), and satisfy Newton's Third Law. But in most cases these provisos don't apply and the body may squash up, bend, stretch, twist, or break. For example, a geologist, using a hammer to apply a force in one direction, may find that a rock cleaves in a completely new direction; a plastic handle, instead of rotating as one body, may buckle, twist, or develop a crack; a girder may bend; a shaft may shear; a vertical column with a load on top may nevertheless explode sideways; and so on.

Newtonian Mechanics is ultimately built up from a framework of individual point-particles and forces, but often appears to avoid the need for internal forces by using whole constructs such as 'rigid body', 'moment of inertia', and so on. Also, the mechanics appears deceptively simple through familiarity (but remember, for example, the sheer variety of formulae needed for moment of inertia depending on whether the body is a sphere, spherical shell, cylindrical shell, rectangular parallelepiped, uniform slender rod, quarter circular rod, right circular cone, elliptic paraboloid, half torus, rectangular tetrahedron, and so on[12]). Many of these techniques were developed by those geniuses of the eighteenth century, notably Leonhard Euler, and Daniel Bernoulli, and even earlier Newton had brilliantly demonstrated that the whole mass of the Earth could be said to act at the Earth's centre (its 'centre-of-mass') as far as the Moon was concerned. These hard-won advances are now taken for granted, and we have forgotten that they contain many implicit assumptions, and that in most realistic cases the system can no longer be modelled as being made up of modular components like 'point-particles', 'forces', 'central forces', 'normal reaction', 'transmissable force', 'forces transmitted infinitely fast', 'inextensible cord', and so on. Then, Newtonian Mechanics fails to solve the problem. This comes as a shock - as we have been led to believe that Newtonian Mechanics fails only when the relativistic or quantum mechanics regimes are reached. Rather, we should realize how amazing it is that such simple elements (forces, point-particles, and accelerations) could ever have accounted correctly for so much.

[12] Meriam J L and Kraige L G, *Statics*, Volume 1, *Engineering Mechanics*, 4th edition, John Wiley & Sons, Inc. (1998) Appendix D, Table D/4.

In summary, Newtonian Mechanics considers the infinity of particles, and the infinity of net forces, one by one, whereas the P of VW (by considering only virtual displacements/motions that are in harmony with the constraints, can thereby ignore these constraints and so) *treats the whole body in one go.*

4.9 Mechanics and geometry

We are witnessing a beguiling melding of geometry and physics. Already in Chapter 2, in Stevin's 'wreath of spheres', we saw how the condition for equilibrium was guaranteed merely by a uniform draping (no gaps or bunching up) of a bead chain over the relevant surfaces; and in this chapter we have seen how geometric conditions (an object moves along a given curve, or on a given surface, and so on) is ultimately due to the presence of internal forces. Careful thinking and examination of many scenarios shows us even more: *the virtual displacements must always be perpendicular to the reaction-forces.* For example, in Feynman's horizontal bar (see Section 4.6), while the displacement of each mass was vertical, in line with the applied force of gravity, yet the reaction-force acted along the length of the bar (preventing the masses from getting closer together or further apart - in other words, maintaining the rigidity of the bar). The displacements and the reaction-force were therefore at right-angles to each other.

A surprisingly tricky example is the case of a sliding block which is pushed across a table-top by a force, say, pushed by your finger (we ignore friction). The displacement of the block is anywhere on the surface whereas the reaction-force acts at right-angles to this surface preventing the block from burrowing down into the table. So far, this makes sense. But, hang on, there is also a reaction against your finger, from the block, and this reaction is *in line* with the block's displacement. The trick is to appreciate that the block's displacement due to the finger-push is an actual, not a virtual, displacement.[13] We can hypothetically freeze the block (switch to a different reference frame) and get rid of the distraction of its actual motion. Then we realize that the finger can't depress the block as if it were so much sponge-cake, as there is a reaction-force of the block against the finger. However, the finger is

[13] We mentioned in Section 4.2, Method 2, how sometimes the actual physical displacement could not be chosen as the virtual displacement.

still allowed, infinitesimally, virtually, to move within the back face of the block, at right-angles to this reaction-force. This is a general result: for any virtual displacement, being 'harmonious' is the same thing as being in a direction perpendicular to the reaction forces.

Optional: It is well known that at equilibrium the potential energy function, V, is not just stationary, it is at a minimum - in most cases. The P of VW cannot determine this, a further investigation is required. Why it is that V is usually at a minimum rather than a maximum will be explained later, in Chapter 6, Section 6.6.

4.10 More examples using the Principle of Virtual Work

(i) The 'black box'

Of all the problems in Meriam and Kraige's textbook, "*Statics*",[14] this simple set-up of connected pushrods is the most marvellous (see Figure 4.2). The pushrods, A and B, can slide in or out of a box. (They are capable only of linear motions but don't need to be positioned in line with each other.) The pushrods are connected together by some series of reversible mechanical devices (racks and pinions, gears, hydraulic pistons, pulleys, levers, and so on) such that when rod A is pushed into the box then rod B gets pushed out, and when rod B is pushed into the box then rod A gets pushed out. We cannot see, hear, or in any other way detect the presence of these internal transmission mechanisms (the box is 'black'), and we must assume that they act without friction or any other dissipation of energy. The question posed is: for every 1.0 unit of inward movement of pushrod A under the action of force, $\mathbf{F}_1$, pushrod B moves outwards from the box 0.25 units against the action of force, $\mathbf{F}_2$. If $\mathbf{F}_1 = 100$ N, determine the magnitude of $\mathbf{F}_2$ for equilibrium (the state where the pushrods don't move). We choose virtual displacements with directions as if pushrod A could drive B out of the box. Then the virtual work done by rod A is $\mathbf{F}_1 \cdot \delta\mathbf{r}_A$ and is positive as the directions of $\mathbf{F}_1$ and $\delta\mathbf{r}_A$ are the same (they are both directed into the box), while the virtual

[14] Meriam J L and Kraige L G, *Statics*, Volume 1, *Engineering Mechanics*, 4th edition, John Wiley & Sons, Inc. (1998) page 449.

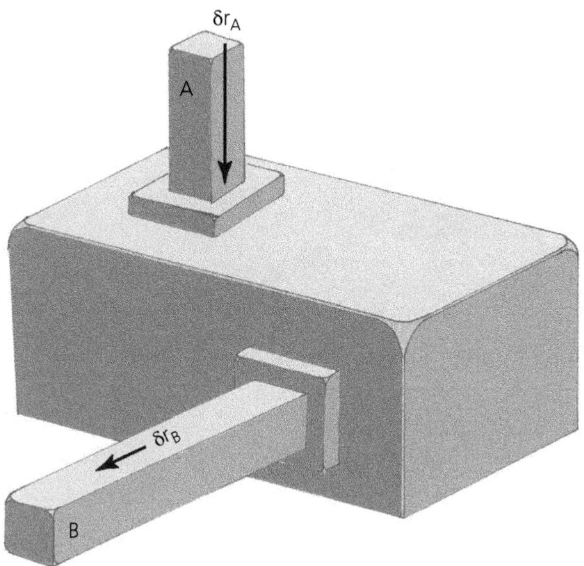

Figure 4.2 The 'black box'
(adapted from Meriam and Kraige, *Statics.*)

work done by rod B is $\mathbf{F}_2 \cdot \delta\mathbf{r}_B$ and is negative as the directions of $\mathbf{F}_2$ and $\delta\mathbf{r}_B$ are opposite ($\delta\mathbf{r}_B$ goes out of the box while $\mathbf{F}_2$ is directed into the box). The total virtual work must be zero for equilibrium, and thus we must have $\mathbf{F}_1 \cdot \delta\mathbf{r}_A = -\mathbf{F}_2 \cdot \delta\mathbf{r}_B$ or $\mathbf{F}_2 = 400.0$ N.

This problem is a marvel because it demonstrates an astounding fact - *the problem is insoluble using Newtonian Mechanics.*[15]

(ii) Knob and slider

This is another kind of 'black box' except that now the angular displacement of a knob, $\delta\theta$, is linked to the linear displacement of a slider, δx. The knob is turned by applying a couple ('turning force') **M** (having units of Newton metres), and then the slider moves against a force, **F** (having units of Newtons). Assuming no dissipative losses

[15] In Newtonian Mechanics we would be compelled to open up the box and assess the forces acting at each and every intersection.

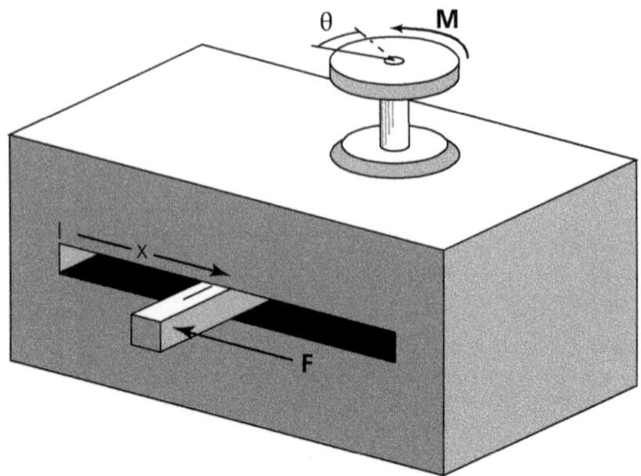

Figure 4.3 Knob and slider
(adapted from Meriam and Kraige, *Statics*, 4th edition.)

within the 'box', for equilibrium the Principle of Virtual Work requires that $\mathbf{M}\delta\theta + \mathbf{F}\delta x = 0$.

(iii) Ladder leaning against a wall

What is the connection between the weight, **W**, and the horizontal friction force, $\boldsymbol{F}_{horiz}$, for equilibrium of the ladder? We consider that the ladder is rigid and of uniform density - so its weight acts at its centre (halfway along its length, L). Also, we assume that the wall is smooth (frictionless) whereas there *is* contact friction from the floor (or else the ladder would fall down). If the top of the ladder has a vertical virtual displacement of δy then the bottom of the ladder will have a corresponding horizontal displacement of δx. This is because, although the displacements are virtual, they are still linked together by the constraint condition that the ladder's length doesn't change. That is, we have $x^2 + y^2 = L$, and so, differentiating, we must have $2x\delta x + 2y\delta y = 0$, which implies $\delta x = -(y/x)\delta y$. Now, by similar triangles, when the top of the ladder falls through δy, the halfway point falls through $\delta y/2$. By the Principle of Virtual Work, the total virtual work due to the weight and the friction force sums to zero: $(W\delta y/2) + (-F_{horiz}\delta x) = 0$. ($W$ and δy are in the same direction and so the work due to the weight is positive; F_{horiz} and δx are in opposite directions and so the work due to friction is negative.) This means that $W = 2F_{horiz}y/x = 2F_{horiz}\tan\theta$, and therefore the

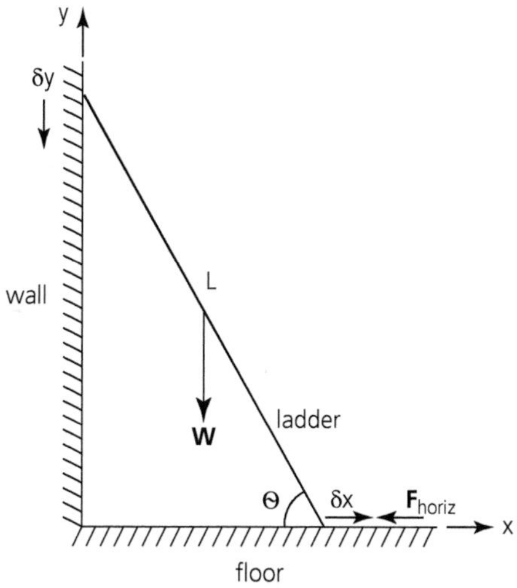

Figure 4.4 Stability of a ladder.

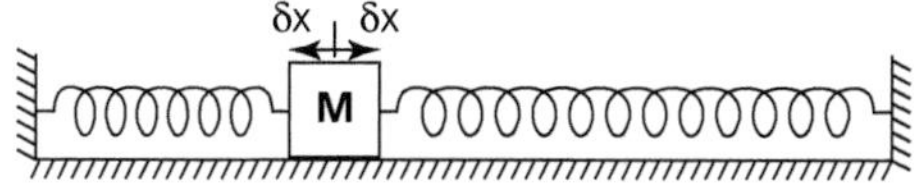

Figure 4.5 Two springs connected to a mass.

final condition for equilibrium of the ladder is: $\theta = \arctan(W/2F_{horiz})$. (The analysis uses magnitudes only.)

Note that as δy and δx are along the wall and floor, respectively, then we have not needed to input the (perpendicular) reaction forces at the wall or against the floor. Note also that the frictional force is that due to *static* friction - so there are no actual displacements of the ladder; the Principle of Virtual Work *must* be used.

(iv) Two identical springs connected to a mass

The Newtonian method easily gives the condition for equilibrium: $\mathbf{F}_1 = -\mathbf{F}_2$. In the method of VW we have: $-\mathbf{F}_1.\delta\mathbf{x} + \mathbf{F}_2. - \delta\mathbf{x} = 0$. The displacement, $\delta\mathbf{x}$, cancels out and then we are left with exactly the same condition as before, $\mathbf{F}_1 = -\mathbf{F}_2$. We see that the virtual work contributions are equal even while one spring is being extended and the other is being compressed - in other words, we learn that *there is no essential difference between extending or compressing a spring* (a fact which is a source of wonderment). It's the same in Newtonian Mechanics (there's no difference between extending or compressing a spring, within the spring's elastic limit) but this similarity now arises for a different reason - because of the postulate that each spring obeys Hooke's Law.

(v) A spring loaded with a weight

How much must a spring be compressed in order to balance the weight of a stone? (Assume the spring's weight can be ignored.) The spring's stiffness is k, its displacement is δx, the stone has mass, m, and the gravitational acceleration is g. We can consider just magnitudes as the problem is in one dimension (vertical). At equilibrium, balancing forces we have: $mg = k\delta x$, whereas balancing energies we have: gravitational energy, $mg\delta x =$ stored spring energy, $\frac{1}{2}k(\delta x)^2$. But this implies that $mg = \frac{1}{2}k\delta x$, which is half what it was before. What has gone wrong? (You may want to cover up the text and think about this for a while before reading on.) The mistake was that, in the energy-analysis, we did not approach the equilibrium state *gradually*. In reality, a displaced spring will not return to equilibrium straight away but will overshoot repeatedly, lose energy by dissipation, and only gradually attain equilibrium. This is curious: we are used to so many textbook problems in which we are told to neglect air-resistance, friction, and so on, but here is a problem where now ignoring dissipative effects leads to the wrong answer. This mistake crept in because we made a false identity between the actual displacement of the spring down to the equilibrium state, and the virtual displacement, δx, of the end of the spring once the equilibrium state had already been achieved.

Both (iv) and (v) are showing that, while the Newtonian method gives us a condition on the exact equilibrium position, the method of VW gives us something more, a condition on this exact position and also on the neighbourhood infinitesimally close to equilibrium.

(vi) Soap bubble

The soap bubble has a higher pressure on the inside (let's call the pressure difference, ΔP) and so it would like to expand, but the surface

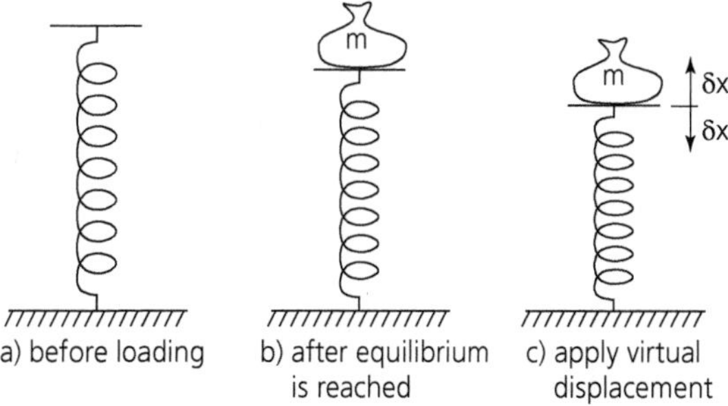

Figure 4.6 Spring loaded with a weight.

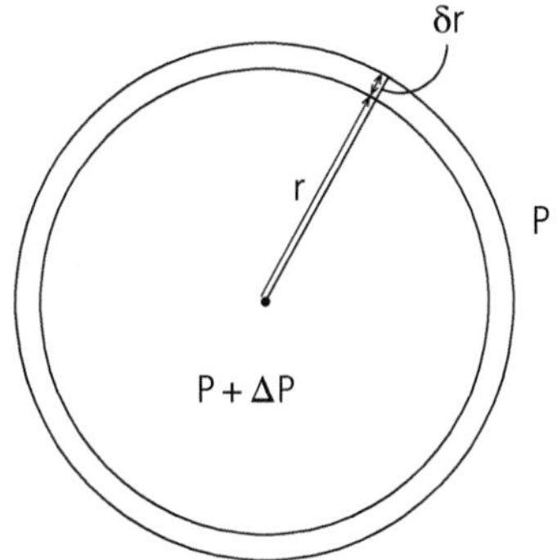

Figure 4.7 Soap bubble.

tension makes it want to contract. Imagine a virtual increase in radius, δr, leading to a virtual volume increase, $\delta V = 4\pi r^2 \delta r$, and a virtual increase in surface area, $\delta A = 8\pi r \delta r$. The virtual work due to the pressure difference is $-\Delta P \delta V = -\Delta P\, 4\pi r^2 \delta r$, and due to the surface tension is, $8\gamma \pi r \delta r$ (where γ is the surface tension of the soap film, a measure of its stored energy per unit area). In fact, there are two surfaces (inner and outer surfaces of the film) and so for equilibrium we must have: $-\Delta P\, 4\pi r^2 \delta r + 2 \times 8\gamma \pi r \delta r = 0$. Therefore, a bubble of radius, r, will be stable so long as $\Delta P = 4\gamma / r$.

(vii) The parallel-plate capacitor

This is adapted from Feynman's *Lectures on Physics*, Vol II, Chapter 8. The spacing between the capacitor plates is increased virtually by δz, and the mechanical virtual work to do this is $F\delta z$ where F is the force between the plates. The virtual change in the capacitor's energy is $\delta U = \frac{1}{2}Q^2\delta(1/C) = -\frac{1}{2C^2}Q^2\delta C$ where C is the capacitance and Q is the stored charge (it is assumed that Q has been kept constant). At equilibrium, the P of VW requires:

$$F\delta z = -\frac{1}{2C^2}Q^2\delta C \tag{4.24}$$

The force is the electrical force of attraction between the plates, but from the P of VW we are learning that F is not affected by the detailed distribution of charge; everything is taken care of by the values of Q and C.

4.11 Lanczos's Postulate A

A marvellous leap forwards in conceptual understanding was brought in by Lanczos, in his re-statement of the Principle of Virtual Work.[16] The usual practice[17] is to define the P of VW as we stated it before in Section 4.4,

$$\sum_i (\mathbf{F}_i^{appl} \cdot \delta\mathbf{r}_i) = 0 \tag{4.25}$$

This sum is the first term in equation (4.19). Lanczos, on the other hand, homes in on the *second* term of equation (4.19), and defines the P of VW as:

$$\sum_i (\mathbf{F}_i^{cons} \cdot \delta\mathbf{r}_i) = 0 \tag{4.26}$$

He then comments that constraint-forces are really the same thing as reaction-forces (we have explained this in Section 4.8) and so re-expresses the P of VW as:

$$\sum_i (\mathbf{F}_i^{reaction} \cdot \delta\mathbf{r}_i) = 0 \tag{4.27}$$

Finally, he puts this in words, and emphasizes that it is a postulate by calling it Postulate A:

[16] Lanczos, page 76.

[17] See for example, Goldstein H, *Classical Mechanics*, Addison-Wesley (1980).

Lanczos's Postulate A

"The virtual work of the forces of reaction is always zero for any virtual displacement which is in harmony with the given kinematic constraints."

But what has been the point of all this wordplay? I surmise that Lanczos's motivations were as follows. The applied forces (squashed spring, finger-push, gravity, electrostatic attraction, and so on) are usually given beforehand, supplied by the problem, and are in the form of scalar mathematical functions. Likewise, the overall kinematical conditions are usually predetermined and in the form of mathematical functions. The reaction forces, by contrast, are consequential - they are, as their name suggests, reactive, responsive, and they are ultimately microscopic in origin, and usually 'polygenic', that is, arising from many sources, and therefore not defined by a scalar mathematical function. Moreover, these reaction forces act 'cooperatively', *as a system*; and so it is from these forces that we obtain the most physical insight. Also, the bald statement of equation (4.27) does not remind us that this is a mathematical procedure that involves 'taking the limit'; and, crucially, that the $\delta \mathbf{r}_i$ must be only in directions allowed by the constraint and kinematic conditions. This essential requirement is brought out by the words 'in harmony with'. Finally, Lanczos is mindful to emphasize that the P of VW is a postulate, and does not come out of Newton's vector-force mechanics. But perhaps the most decisive reason for Lanczos to put the Principle of Virtual Work into words is that, as Postulate A, it will lead into the whole of the rest of mechanics (the variational method as applied to both statics and dynamics), and is therefore of astounding universality:

"Postulate A is actually the *only* postulate of analytical mechanics, and is thus of fundamental importance."[18]

4.12 Concluding Remarks

There are some questions that we're not allowed to ask:

Why does the P of VW work? Answer - we don't know, it just does.

How do the forces-of-reaction act cooperatively? Answer - we don't know, they just do.

[18] Lanczos, page 76.

How can a test in a fictitious space tell us anything about the real world? We don't know, but it happens also in countless other circumstances. Consider the statistic, "an average family of 2.2 children"; evidently there can never be a family of this size but the statistic may still serve as a useful criterion. Also consider the example of the hypothetical money-transfers postulated by a financier in a 'test of portfolio value'.[19] The money-transfers are subject to certain fees, interest rates, and exchange rates, but in order to actually buy and sell shares, or transfer money between accounts, some actual time is required. During this time, the financial conditions may change - and so the 'test of portfolio value' occurs in a *virtual* space.

But this is only half the answer. The utility of using a virtual space is that it can provide an absolute benchmark in a way that no actual physical space could ever do. This absoluteness and invariance of the results in the abstract test space is what comes about in the branch of mathematics known as 'differential geometry' (which used to be called 'absolute geometry'). The results are the same irrespective of what coordinates have been chosen or how the system has been modelled. We shall find that this remarkable invariance is a feature of the whole of Variational Mechanics (the P of VW, Principle of Least Action, Lagrangian Mechanics, and Hamiltonian Mechanics).

An enormous advantage of using the Principle of Virtual Work, as shown in the Examples, Section 4.10, is that we do not need to know the internal forces, the reaction forces, or the forces of constraint. However, a disadvantage of the P of VW is that when, on occasion, the applied force is not given in the form of a scalar mathematical function (for example, the applied force actually *is* a friction-force) then mathematical methods - the Variational Mechanics - cannot be used. On the other hand, as Lanczos notes,[20] given that ultimately all dissipative forces are microscopic in origin, with an appropriate mathematical form derivable from quantum theory, then ultimately it should be possible to apply the variational methods across the whole of physics.

Now the fact that the P of VW yields absolute answers - that's mathematics, but the fact that it's *work* that's minimized as opposed to some other quantity - that's physics. The innumerable cases where the P of

19 Murray Peake, private communication.

20 Lanczos, Introduction, page xxv.

VW succeeds and Newtonian Mechanics fails shows us that it is indeed work rather than force that is pre-eminent. In problems of exceptional simplicity, described in rectangular coordinates, it would be a mistake not to use Newtonian methods and benefit from the physically intuitive forces, but we must always remember that "force is a secondary quantity derived from ... work."[21] As Lanczos puts it:

> "we are inclined to believe that force is something primitive and irreducible, [however] ... it is not the force but the *work* done by the force which is of primary importance."[22]

This is the main reason for using the Principle of Virtual Work: it is abstract, intuitively opaque, and mathematically complicated, but it is the way the physical world really is.

[21] Lanczos, page 27.

[22] Lanczos, page 27, edited and with italics added.

5

D'Alembert's Principle

5.1 Introduction

In the Principle of Virtual Work (the P of VW, Chapter 4) we have a method for analysing the condition known as static equilibrium or 'statics'. In this chapter we widen the scope to include 'dynamics' - mechanical scenarios in which there is motion between the parts of a system. Motion appears antithetical to statics and yet, in d'Alembert's Principle, we will discover a brilliant strategy for treating these dynamic cases; they will be treated, once again, as problems of 'equilibrium'. In essence, d'Alembert's Principle proclaims that the very motions are exactly such as to bring the system, instant by instant, back to 'equilibrium' - an equilibrium of a new, *dynamic* kind. The Principle relies on one radical new idea, d'Alembert's "stroke of genius"[1]: it is that a mass-in-accelerated-motion counts as a force. Previously, we had only Newtonian-type forces (cohesive, tensional, gravitational, electrical, magnetic), and these are all forces emanating from some source or other (a gravitating mass, an electric charge, a magnet, and so on). Now, with d'Alembert, we introduce a totally new type of force - a mass, merely by virtue of its motion, can act as a force. With hindsight, we see that this must be so - if 'motions' can counteract forces, then 'motions' must *be* forces. D'Alembert's programme is then to add this new category of force to the familiar applied- and constraint-forces, and apply the Principle of Virtual Work in the usual way.

D'Alembert's new force, called the force-of-inertia, or 'inertial force', occurs wherever there is a mass-in-accelerated-motion. A mass that is moving but not accelerating does not count as a force, and a geometric point that is accelerating but is massless (such as the geometric point of intersection between two accelerating rods) does not count as a force. Also, it is of no consequence to ask how the mass-in-accelerated-motion

[1] Lanczos, page 88.

The Lazy Universe. Jennifer Coopersmith, Oxford University Press (2017).
 DOI 10.1093/acprof:oso/9780198743040.001.0001

gets to be accelerating - it could be subject to a force (such as an applied force), or it could be viewed from a reference frame that is accelerating, or a mixture of these. Howsoever it originates, the inertial force - let's call it **I** - has all the usual attributes of a force: it has magnitude and direction, and it adds and 'multiplies' in the usual vectorial ways. The inertial force may therefore be added to the other forces, and then the Principle of Virtual Work may be employed.

Let us see how this happens. We consider a mechanical system composed of N particles ($i = 1$ to N) of mass, m_i, position vector, $\mathbf{r}_i$, subject to applied forces, $\mathbf{F}_i^{appl}$, and/or constraint forces, $\mathbf{F}_i^{cons}$. From Section 4.8, Chapter 4, we remember that the applied- and constraint-forces have very different provenances: the applied forces are described by mathematical functions, specified beforehand in the given problem (for example, gravity follows an 'inverse square law'), whereas the constraint forces (also called internal forces or reaction forces) are generally unknown and not expressible as a mathematical function (what is the force between particles in a taut cord, or in a rigid body, and so on?). Despite these different origins, we add together the applied- and constraint-forces acting on a given particle, and call this sum $\mathbf{F}_i^{net}$. Now, in the usual way (that is, in accordance with Newton's Second Law of Motion), $\mathbf{F}_i^{net}$ acting on the mass, m_i, causes it to accelerate, and so we must have $\mathbf{F}_i^{net} = m_i\mathbf{a}_i$. But, by d'Alembert's "stroke of genius", this massy acceleration, in and of itself, constitutes a force. Even more than that, by postulate it constitutes a *reactive* force, a force that always *opposes* $\mathbf{F}_i^{net}$. Therefore, we define the inertial force, $\mathbf{I}_i$, as being in the *reverse* direction to $\mathbf{F}_i^{net}$, that is, in the *reverse* direction to the acceleration:

$$\text{Inertial force defined as:} \qquad \mathbf{I}_i = -\mathbf{F}_i^{net} = -m_i\mathbf{a}_i \tag{5.1}$$

Now the total force at i is the sum of $\mathbf{I}_i$ and $\mathbf{F}_i^{net}$; we call this the 'effective force', $\mathbf{F}_i^{eff}$:

$$\mathbf{F}_i^{eff} = \mathbf{F}_i^{net} + \mathbf{I}_i \qquad \text{for particle } i \tag{5.2}$$

Thus for N particles we end up with a system of forces, $\mathbf{F}_1^{eff}, \mathbf{F}_2^{eff}, \ldots, \mathbf{F}_N^{eff}$, and, by d'Alembert, we seek an equilibrium condition between these forces, even for this dynamics (non-statics) scenario.

This seems very nice until we realize with some surprise that, by equations (5.1) and (5.2), the $\mathbf{F}_i^{eff}$s are all identically zero: $\mathbf{F}_1^{eff} = 0$,

$\mathbf{F}_2^{eff} = 0, \ldots, \mathbf{F}_N^{eff} = 0$. Nevertheless, we are free to 'multiply' each zero $\mathbf{F}_i^{eff}$ by a virtual displacement, $\delta\mathbf{r}_i$, and we can be assured that each 'product', $\mathbf{F}_i^{eff} \cdot \delta\mathbf{r}_i$, will also be identically zero. We can then sum these zeros together and, of course, end up with zero:

$$\sum_{i=1}^{N} \mathbf{F}_i^{eff} \cdot \delta\mathbf{r}_i = 0 \tag{5.3}$$

While the correctness of this equation is not in doubt, the utility of it is unclear. For one thing, we still don't know what the $\mathbf{F}_i^{eff}$ actually are (we know that they're identically zero, but we don't know *what* is being equated to zero); we have been supplied with the applied forces, and we seek to determine the accelerations (the ones that will lead to 'equilibrium'), but we are stymied by our lack of knowledge of the constraint-forces. Nevertheless, we press on, and try decomposing each $\mathbf{F}_i^{eff}$ into its constituent force-types, that is, applied, constraint, and inertial forces. Equation (5.3) then becomes:

$$\sum_{i=1}^{N} (\mathbf{F}_i^{appl} + \mathbf{F}_i^{cons} + \mathbf{I}_i) \cdot \delta\mathbf{r}_i = 0 \tag{5.4}$$

At this stage we do some rearranging of the three terms within the bracket - by exploiting the well-known rules of arithmetic. Thus, using the commutative and associative laws, we may group the threesome into $(\mathbf{F}_i^{appl} + \mathbf{F}_i^{cons}) + \mathbf{I}_i$, $\mathbf{F}_i^{appl} + (\mathbf{F}_i^{cons} + \mathbf{I}_i)$, or $(\mathbf{F}_i^{appl} + \mathbf{I}_i) + \mathbf{F}_i^{cons}$. Choosing the last one and using the distributive law, (5.4) becomes:

$$\sum_{i=1}^{N} (\mathbf{F}_i^{appl} + \mathbf{I}_i) \cdot \delta\mathbf{r}_i + \sum_{i=1}^{N} (\mathbf{F}_i^{cons}) \cdot \delta\mathbf{r}_i = 0 \tag{5.5}$$

At last, we have made some progress - we have herded all the troublesome (that is, unknown) constraint-forces into one group. The way forwards is now clear: *provided* we insist that all the virtual displacements are in conformity with (in 'harmony' with) these constraint-forces, then we can be assured that the constraint-forces will do no virtual work. (This has been explained in Section 4.9, Chapter 4.)

The constraint-forces, separately from all the other forces, will then make up a sub-system in equilibrium:

$$\sum_{i=1}^{N}(\mathbf{F}_i^{cons}) \cdot \delta\mathbf{r}_i = 0 \tag{5.6}$$

This saves us - it is simply unnecessary to know what the constraint forces are (their contribution to virtual work is zero). But, as we still require that the total virtual work from *all* sources equals zero (equation (5.5)), then it must be that the applied and inertial forces, taken together, also form a sub-system in equilibrium (their total virtual work must also be zero). This, finally, is what is conventionally referred to as d'Alembert's Principle:

D'Alembert's Principle, standard formulation

$$\sum_{i=1}^{N}(\mathbf{F}_i^{appl} + \mathbf{I}_i) \cdot \delta\mathbf{r}_i = \sum_{i=1}^{N}(\mathbf{F}_i^{appl} - m_i\mathbf{a}_i) \cdot \delta\mathbf{r}_i = 0 \tag{5.7}$$

Note that although the sum goes from 1 to N, there is not necessarily an applied force at every i (there could be an accelerating mass-point even where the applied force happens to be zero).

There is yet one more change in nomenclature that we can introduce. The bracket $(\mathbf{F}_i^{appl} - m_i\mathbf{a}_i)$ is equivalent to one resultant force at i, and we can name this resultant force as we choose. Let us call it the 'dynamics force', $\mathbf{F}_i^{dyn}$, as we reach an equilibrium of a new *dynamic* kind, that is, an equilibrium where accelerations occur. We can then rewrite (5.7) finally as:

D'Alembert's Principle

$$\sum_{i=1}^{N}\mathbf{F}_i^{dyn} \cdot \delta\mathbf{r}_i = 0 \tag{5.8}$$

This is exactly the same as the (standard definition of) the Principle of Virtual Work, see (4.21), as long as we pay no attention to the different naming conventions ($\mathbf{F}_i^{appl}$ occurs in the P of VW, $\mathbf{F}_i^{dyn}$ occurs in d'Alembert's Principle) - but then what's in a name?

Et voilà - dynamics has been reduced to statics. Well, almost; there is a difference between statics and dynamics, and this goes deeper than the mere statement of the Principle (definition (4.21) as opposed to definition (5.8)); the difference lies entirely in the solutions to (4.21) and to (5.8), in other words, the final equations are *not* the same. In statics we end up with *statical* relations (a weight might have to be located at a certain distance along a beam, a taut cord located at a certain angle, a weight squashes a spring to just such an extent, and so on) whereas in dynamics the Principle determines the *accelerations* of particles, and then if we wish to know how the positions of these particles change with time we will have to go on to solve the appropriate (2nd order) *differential equations of motion.*

Despite our success in combining both statics and dynamics into one Principle, there is much that is perplexing and disturbing about our 'derivation'[2] (equations (5.1) to (5.8): naming, renaming, and shuffling packets of zero. . .). We have the feeling of whipping up a soufflé with no eggs, and, to compensate for the absence of eggs, whipping faster and faster. Before we return to this, it will help to examine a worked example.

5.2 A worked example

We provide a high-school physics example[3] to show how d'Alembert's Principle works in practice. A mass, m_1, slides across a table-top and is joined, by a rope going over a pulley, to another mass, m_2, which is falling under gravity (Figure 5.1 a). We make all the usual idealizations: m_1 slides without friction, the rope is taut and inextensible, there is no friction at the pulley, and the masses of the rope and pulley can be ignored. In fact, we can idealize the whole experimental arrangement as a 'black box' with m_1 approaching, m_2 receding, plus appropriate forces (Figure 5.1 b) (cf. problem (i) Section 4.10, Chapter 4).

The virtual displacements of m_1 and m_2 are along the length of the rope (they're 'harmonious'). Also, as the rope-length is fixed, the motions of m_1 and m_2 not independent; we must have $\delta\mathbf{r}_2 = \delta\mathbf{r}_1$ and $\mathbf{a}_2 = -\mathbf{a}_1$. We take all directions into (out of) the 'box' as negative (positive). From (5.7) we obtain: $(\mathbf{F}_1^{appl} + \mathbf{I}_1)\cdot\delta\mathbf{r}_1 + (\mathbf{F}_2^{appl} + \mathbf{I}_2)\cdot\delta\mathbf{r}_2 = 0$. Setting $\mathbf{F}_1^{appl} = 0, \mathbf{F}_2^{appl} = \mathbf{W}_2 = m_2\mathbf{g}, \delta\mathbf{r}_1 = -\delta\mathbf{r}_2$, and $\mathbf{a} = \mathbf{a}_2 = -\mathbf{a}_1$, and remembering that $\mathbf{I}_i = -m_i\mathbf{a}_i$ for all i, we obtain: $(0 + m_1\mathbf{a})\cdot(-\delta\mathbf{r}_2) + (m_2\mathbf{g} - m_2\mathbf{a})\cdot\delta\mathbf{r}_2 = 0$.

[2] The definition (5.8) was given so as to emphasis the identicality of d'Alembert's Principle with the Principle of Virtual Work. From now on, we shall always use the definition of d'Alembert's Principle as given in (5.7).

[3] Year 12 VCE physics homework, Bendigo Senior Secondary College, 2013.

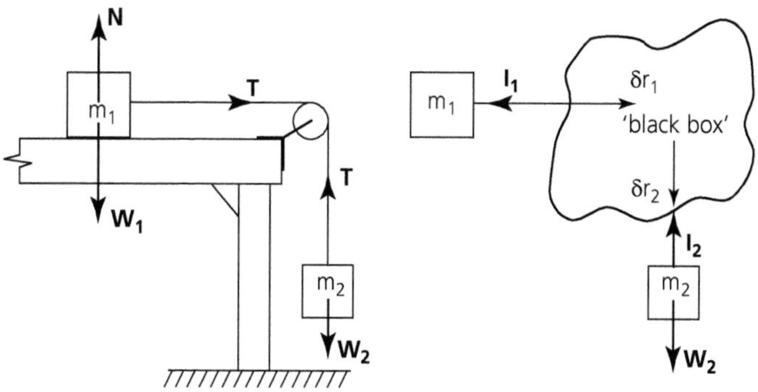

Figure 5.1 Connected masses a) "Half Atwood" version b) "Black box" version.

Cancelling $\delta\mathbf{r}_2$ and rearranging, we arrive at the standard solution: $\mathbf{a} = m_2\mathbf{g}/(m_1 + m_2)$. Note that we have not needed to calculate the tension of the rope as this is a constraint force and does no virtual work (but in the method of Newtonian Mechanics we would need to determine the tension, or know this beforehand). Even more interesting is the role played by the mass, m_1. It undergoes an acceleration and therefore introduces an inertial force, $\mathbf{I}_1$ (in the reverse direction to this acceleration), but this inertial force does not occur in tandem with any applied force. Rather, $\mathbf{I}_1$, exists because of the mere fact of m_1's acceleration, and we end up with the curious result that $\mathbf{I}_1$ is the sole influence of m_1 on m_2; the weight of m_1, and its force-of-attachment to the rope, are both irrelevant as far as m_2 is concerned (provided that m_1 does indeed remain attached to the rope[4]). This is evidence that the inertial force truly exists, and is just like any other force.

5.3 Against intuition?

We return now to the question that we left dangling at the end of Section 5.1, and to other counter-intuitive aspects of d'Alembert's Principle.

If the 'derivation' of d'Alembert's Principle (equations (5.1) through to (5.8)) arose out of nothing more than the renaming and regrouping of packets of zero then this would indeed be paradoxical; however, the paradox is resolved when we appreciate that *new knowledge* has entered

[4] As with all methods in the Variational Mechanics, d'Alembert's Principle only works for sufficiently small virtual displacements leading to sufficiently small effects.

into the problem. (This is reminiscent of a well-known 'paradox' in probability - the tale of the three caskets, or the Monty Hall problem.[5]) Specifically, in equation (5.6) we have introduced an extra postulate (asserting that the virtual work of the constraint forces is zero), and this brings in the new knowledge that the motions of the mass-points are dependent on each other, and may even be tied together into 'super-structures' (wheel, lever, inclined plane, and so on). Extra knowledge has also been brought in by the requirement that the virtual displacements can't be in arbitrary directions but must be in harmony with these super-structures (and with any other kinematic conditions that may apply).

What about if there are no constraints, does d'Alembert's method then reduce to Newtonian Mechanics? Not necessarily - although it is true that the simpler the system is, the less is the difference between the methods, and the less apparent is the advantage of using d'Alembert's Principle. Consider the very simplest case: one particle, one applied force, and no constraints. We then have $\mathbf{F} = m\mathbf{a}$ (from Newton's Second Law of Motion), and $\mathbf{F} - m\mathbf{a} = 0$ (from d'Alembert's Principle, equation (5.7), after superscripts and subscripts have dropped away, and $\delta\mathbf{r}$ cancels out). It seems as if d'Alembert's Principle has offered nothing more than a trite rearrangement of Newton's Second Law. However Newton and d'Alembert are really posing utterly different questions. Newton asks - what is the motion of a particle given that the applied force is $\mathbf{F}$? D'Alembert asks - what is the condition for equilibrium when there is a collection of forces? He finds that the condition is $\sum(\text{forces}) = 0$ where it just so happens that the forces are '$\mathbf{F}$' and '$-m\mathbf{a}$'. The wording is telling: Newton refers to 'a particle', in other words, to *one* particle; d'Alembert refers to a '*collection* of forces', in other words to a whole *system*. But the

[5] An ancient king needs to find a husband for the princess. The king has three identical caskets. He hides treasure in one of them and leaves the other two empty. Furthermore, he makes sure that no one but he knows which casket contains the treasure. In one go, the successful suitor must choose the casket containing the treasure. One day a suitor whom the King especially likes arrives at the palace. This suitor chooses a casket but before he gets to it the King quickly opens one of the other caskets, revealing - emptiness. The King then gives the suitor the option of changing his original choice. The 'paradoxical' question is: in order to maximize his chance of winning the princess's hand in marriage, should the suitor change his choice or stay with his initial choice? This seems like a 'no-brainer' (of course the king's intervention won't change the odds) until we realize that the king is introducing *extra knowledge* into the problem...

differences go even deeper than this: Newton's Second Law is an equation in vector algebra, while d'Alembert's $\sum(\text{forces}) = 0$ is a shorthand for a *principle* in the variational calculus: it is only true 'in the limit', and the virtual variations must be chosen wisely.

The elemental system, 'one particle, one force', is so simple to state and easy to visualize that it leads us to suppose that scenarios of any degree of complexity can always be built up out of these basic elements. But this is not so, and the pathological simplicity has led us to false expectations. It is an example of the dictum "exceptionally simple cases make bad physical intuition"[6] (in contrast to the lawyers' dictum "hard cases make bad law").

There are other counter-intuitive aspects of d'Alembert's Principle. The fact that the zero condition in (5.7) involves a *sum* allows for two surprising possibilities. First, it is possible to have some terms where $\mathbf{F}_i^{appl} \neq m_i\mathbf{a}_i$. There is no contradiction with Newton's Second Law as the $m_i\mathbf{a}_i$ are consequent upon the *net* force, $\mathbf{F}_i^{net}$, they are not generated by just the applied force, $\mathbf{F}_i^{appl}$. (Typically, the applied forces will exist at just a few mass-points, whereas the inertial forces will occur at every i, that is, wherever there is a mass in accelerated motion.) Second, and more alarming, it may occur that the interaction force (a genre of constraint force) between one particle and another particle is not 'equal and opposite', that is, there is nothing to prohibit $\mathbf{F}_{ij}^{int} \neq \mathbf{F}_{ji}^{int}$. This means that Newton's Third Law of Motion has failed. There is ample experimental confirmation of this[7] - but note that the failure occurs only at this 'microscopic' level, the level of interacting particles: overall, at the 'whole-system' level, Newton's Third Law does still always apply.

There is yet one more aspect that is against our intuition. We have said that $\mathbf{F}_i^{net}$ leads to the massy accelerations, $m_i\mathbf{a}_i$, but this doesn't rule out the possibility of these $m_i\mathbf{a}_i$ being generated in a totally different way (even when $\mathbf{F}_i^{net}$ is zero): the massy accelerations can occur merely as a consequence of an accelerating reference frame. But this then means that inertial *forces* can occur merely as a consequence of accelerations. This is very non-Newtonian and so will bring in some very non-Newtonian outcomes which we shall now investigate.

[6] The author invented this dictum and used it first in a slightly different context; see Coopersmith, *EtSC*, Chapter 18.

[7] Famously, this occurs in electrodynamics, such as the interaction between two electrons, with one electron approaching at right angles to the path of the other electron.

5.4 'Fictitious forces'

5.4.1 Introduction and examples

Let us generalize from the 'pathologically' simple scenario of just one particle and one force, and include the possibility of many particles, and of constraint or internal forces - is our system now of the most general kind? No, there is yet one more generalization we can make: we can remove the Newtonian prohibition against accelerating reference frames. We give a qualitative account of the consequences of allowing such accelerating reference frames (see Figures 5.2a to 5.2e).

Consider a system seen simultaneously from a reference frame that has no acceleration, REF, and from a reference frame that is accelerating, REF^{acc}. We will find that particles that were stationary in REF will be accelerating in REF^{acc}; and, *per contra*, there may be particles that were accelerating in REF and are no longer accelerating in REF^{acc}. Now, according to d'Alembert, a mass that accelerates constitutes a force. So, merely dependent on the reference frame, *a force could come into existence which didn't exist beforehand.* The consequences are not innocent, a force is not a tame thing (applied to a person, it could make them feel nauseous, lose their balance, or tear them limb from limb). Such forces have in the past been called 'fictitious forces' - a misnomer, as these forces are every bit as real as applied- and constraint-forces, and their existence has been corroborated in countless experiments. (For example, the 'fictitious forces' associated with a rotating reference frame are the well-known centrifugal, and Coriolis forces[8]).

Let's look at some examples. Imagine that you are in a hotel room (a one-room cabin) with the window-shutters closed, in an armchair against the wall (Figure 5.2a). Unbeknownst to you, some prankster has mounted the whole cabin, at its centre, onto a strong bearing with rotary motor attached. The prankster turns the motor on and the room starts to spin (Figure 5.2b). Let's imagine that it spins exceptionally smoothly and quietly. What will you notice? You may feel a bit sick and you'll feel yourself being pushed against the chair-back.[9] The hanging lamp near your chair gets tugged towards the outside wall. Also, some marbles on the floor suddenly start to move towards the edges

[8] There are also the less well-known 'Euler forces' associated with a frame rotating non-uniformly, see Lanczos, page 105.

[9] The faster the spin-rate, the greater your distance from the spin-axis, and the greater your mass, the more strongly you'll feel pinned to the chair.

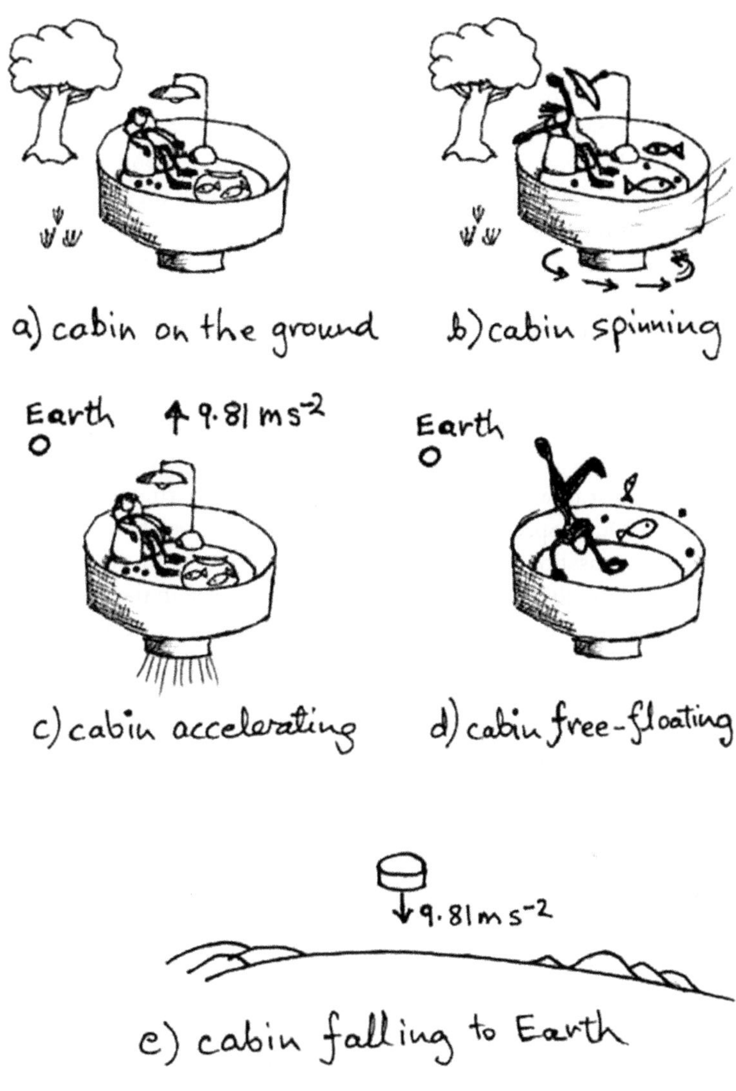

Figure 5.2 'Fictitious Forces', examples.

of the room. They start rolling, radially outwards, but they also feel, in addition, a sideways push, and so pursue a curved path towards the walls.

You exchange stiff words with the management, switch to a new cabin, and promptly fall into a deep sleep. When you wake up later, you are relieved to find that everything is as usual, and you amuse yourself playing with the marbles, having a game of darts, and watching the goldfish in the goldfish-bowl. Everything appears utterly normal until you open the shutters and see - nothing. The view has disappeared and all you can see is a blue planet, far away. Room service (!?) informs you that you are now in Outer Space, on a rocket travelling dead straight and accelerating smoothly at 9.81 metres per second per second, (the engines are under the floor), and, not to worry, you have all the necessary oxygen, cabin pressure, supplies, shielding, and such like.

Your adventure is not over yet: you suddenly notice that every loose object in the room starts to accelerate linearly and gently towards the ceiling. This lasts for some minutes, gradually tailing off, until you notice that you feel weightless and see some marbles hanging motionless in mid-air (you are still on your chair, which is bolted to the floor, but you must now hold on to it). Room service informs you that you are still in Outer Space, but the rocket engines have been turned off. You enjoy the sensation of weightlessness, pushing off from the walls and doing somersaults with ease. Exhausted, you doze off in the chair again, (buckling yourself in). When you awake, you continue to feel weightless and are just about to indulge in some more acrobatics, when you look out the window and notice with horror that the Earth is close by and you are plummeting down towards it (that is, you're in free-fall).

Before you crash to your death, you just have time to ponder on the wisdom of d'Alembert: from within your room you feel only the *combined* force, $(\mathbf{F}^{appl} + \mathbf{I})$, and there is no way for you to distinguish between the effects due to an externally applied force and the effects due to your acceleration - in other words, $\mathbf{F}^{appl}$ *and* $\mathbf{I}$ *are completely indistinguishable.*

Now, if it so happens that the acceleration of your reference frame satisfies two conditions - it is in a straight line, and it's magnitude doesn't change - then the 'fictitious' force in this special case is indistinguishable from a particular kind of external force - the one known as gravity.[10] This was Einstein's celebrated Equivalence Principle. D'Alembert's Principle is more general inasmuch as the external force doesn't have to be gravity, but the Equivalence Principle is more general inasmuch as it applies not only to mechanics (bodies in motion) but to any physical effect whatsoever (heating, electricity, radioactivity, chemical action, and so on). In other words, all these effects will be unaltered when

[10] (for a sufficiently small room and short timescale).

viewed from a reference frame that has zero acceleration and includes a gravitational field, or has a constant linear acceleration and does not include a gravitational field. Einstein subsequently generalized the Equivalence Principle in his landmark Principle of General Relativity: in a closed system, no physics experiment, of any kind, can distinguish between reference frames having *any* kind of motion (zero acceleration, uniform acceleration, or non-uniform acceleration[11]).

This is all completely contrary to Newtonian Mechanics. Newton makes the distinction between a free body (there are no forces acting on it), and one that is subject to a force. By contrast, d'Alembert, Einstein, and modern physics, all make a distinction between a constrained body (moving only within a given surface, tied to a cord, and so on), and an unconstrained body - whether or not there may additionally be external influences. The importance of force is now downplayed, and we sometimes find (for example, in modern field theory) that it disappears altogether.

5.4.2 *Critique, and some historic examples*

One may try the objection that scenarios (a) and (c), and again (d) and (e), (Figure 5.2) do not have to yield the same outcomes as they are not identical (rocket engines are on and off, the Earth is near or far). This is true but, first, d'Alembert's Principle, and the Principle of General Relativity, do not deny that **F** and **I** can be distinguished if you look outside the cabin (that is, look outside the system, look at 'the view'), and second, these Principles don't prohibit identical outcomes for different scenarios, they only prohibit different outcomes for identical scenarios.

Let's nevertheless, make things simpler by considering just one[12] scenario (which must, presumably, be identical to itself!) as seen from different view-points. For example, consider the Pluto-Charon system (and make the approximation that Pluto and Charon are perfectly rigid bodies travelling in circular orbits), and then site the origin of the reference frame: (a) on Pluto, (b) on Charon, and (c) at the centre-of-mass of the whole system (that is, on the line joining Pluto and Charon, just slightly outside Pluto's body). Surely, as there is only one actual

[11] Again, provided that the experiment is only carried out over 'small' time- and space-scales.

[12] A cabin that is left alone, rotated by an engine, or fired into space by a rocket, all count as different scenarios.

scenario, then d'Alembert's Principle will guarantee the same 'outcomes' for each view-point? Yes, it does, but we can hardly say that the observed motions are the same in (a), (b), or (c). Now here's a strange thing. Earlier on (in the spinning cabin) we showed that there were 'fictitious forces' as a result of rotary motion (you were pinned to your chair, you felt a bit sick, the lamp no longer hung vertically down, and marbles started moving) but now, despite the fact that both Pluto and Charon are orbiting and spinning, yet there are no centrifugal or other 'fictitious' forces, no evidence of rotation whatsoever (apart from looking outside the system at 'the view', that is, the distant stars). How does this curious result arise? It arises because of the simplifying assumption - that Pluto and Charon are perfectly rigid bodies. In the cabin this assumption has not been made (humans are squishy, marbles are not stuck to the floor, the hanging lamp can swing). In fact, it is *only* through these non-rigid attributes that rotations can be noticed.

Another example is that of a bucket, spinning around its vertical axis of symmetry. The bucket would give no evidence of its acceleration unless it was filled with something non-rigid, like water. This is the famous example of Newton's rotating-bucket experiment. Newton noted that, after spinning for a while, the surface of the water becomes curved, with the outer edges higher up than the level in the middle (Figure 5.3). Newton took this curved water-surface as a demonstration that the bucket really is spinning (relative to an unobservable but absolutely stationary background Space).

Ernst Mach (nineteenth century physicist and philosopher) had an intriguing alternative idea. He saw that a) was not the true inverse of b) (see Figure 5.3), as the bucket of water was not, in reality, surrounded by empty space but by the distant stars. If, in b), the distant stars were stationary, then, in the true inverse of b) (say, c)) they should be rotating at the same rate[13] but in the opposite direction to the bucket's rotation. As the background system of stars would then be different in b) and c), then this could account for foreground differences, like the different shape of the water's surface. In effect, Mach asserted that the stars actually caused the water-surface to curve.

Nowadays we reject Newton's absolutes (force, acceleration, Space), and we also reject Mach's hypothesis. We can never carry out Mach's

[13] We have ignored the spin of the Earth, but this can be accounted for and doesn't change the argument.

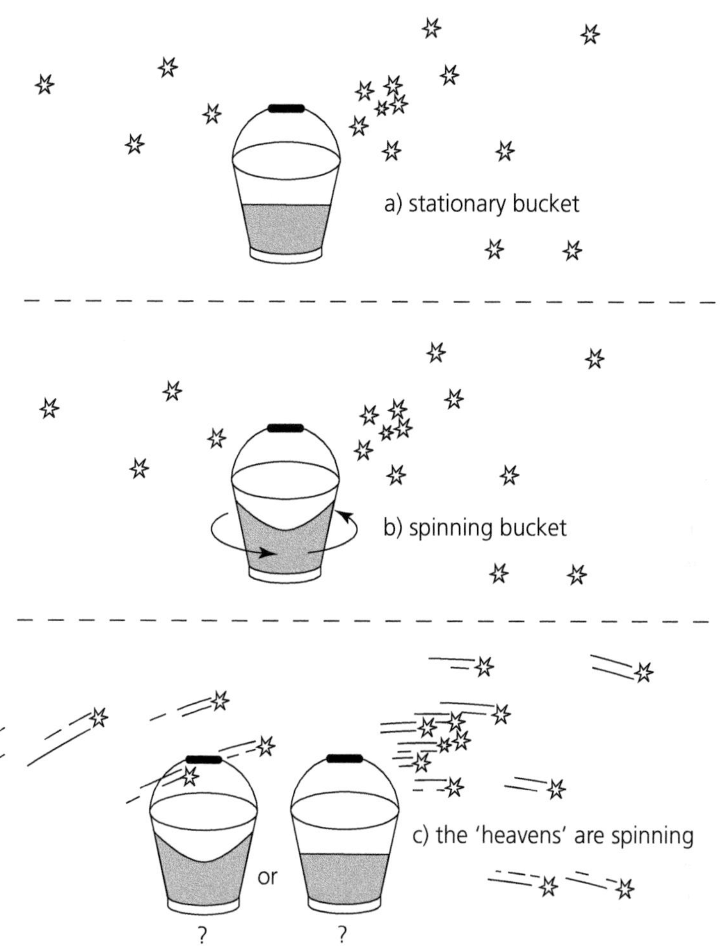

Figure 5.3 Newton's Bucket.

hypothetical experiment of spinning or removing the heavens, and so we can never confirm or deny Mach; yet, his hypothesis is completely against the spirit of d'Alembert's Principle (and against the spirit of all variational mechanics, and Einstein's Theory of General Relativity). While both Mach and d'Alembert agree that what is important is the system, the whole system, and nothing but the system, yet, in d'Alembert's Principle, the system does not extend to infinity, it only encompasses *local* effects and only makes *local* claims. This is more humble and more philosophically correct ('pc'). In the variational mechanics, which includes d'Alembert's Principle, the curving of the water-surface is due only to *local* interactions (neighbouring molecules), and *local* motions (of water molecules with respect to the centre of the bucket - a small distance). All this *can* be tested experimentally. Now, the bucket radius is smaller than the distance to the stars, yes, but how small is 'local'? Ah, that depends. . . [14]

5.5 D'Alembert's Principle and energy

There is one matter concerning d'Alembert's Principle that we haven't mentioned so far - the question of solvability, that is, can mathematical solutions to (5.7) be obtained? The applied forces are usually specified beforehand, and given in functional form (Hooke's Law for a spring, Coulomb's Law for electrostatic attraction, and so on) whereas the inertial forces (reversed accelerations of the mass-points) are not usually in functional form.[15] This is because these accelerations have arisen from *multiple* influences - constraint forces as well as applied forces. An excellent demonstration of this is seen in the system of the roulette wheel and ball: here there are two applied forces - gravity and the push given to spin the wheel - but the resulting acceleration of the roulette-ball is all over the place - 'randomized' - due to the complexity of the multiple constraint forces (the ridged and sloping surface of the spinning roulette wheel).

Now there is one special circumstance in which the (reversed) accelerations *can* be expressed in functional form. Remember that d'Alembert's Principle uses virtual displacements that are of our choosing (subject to any constraint and kinematical conditions): surely, we can choose the $\delta\mathbf{r}_i$ to coincide with the actual displacements, $d\mathbf{r}_i$, which really do occur? This would resolve our difficulties as the $d\mathbf{r}_i$ are 'perfect differentials' and therefore amenable to mathematical solution. But, strange as it may seem, the actual displacements are not always permissable as a stand-in for the virtual displacements. This is because the

[14] That depends on the precision of the measuring apparatus, and the rapidity of change in the external forces and other conditions - see Chapters 3 and 4.

[15] The inertial accelerations are not the familiar 'rate of change of velocity', $d\mathbf{v}/dt$. They are tiny incremental vectors, in the correct units, but they do not have a functional form (cf. the guinea-pig nibbling the biscuit (Section 3.7.2 c)).

actual displacements, $d\mathbf{r}_i$, always take a finite time, dt, and during this finite time-interval the external conditions (applied forces, constraint forces, kinematical conditions, other external conditions) may change. However, in the one special circumstance that *all the external conditions are constant*, independent of time, then there will be no contradiction in choosing the $\delta\mathbf{r}_i$ to be the same as the $d\mathbf{r}_i$.

In this special circumstance in which everything is in functional form, and everything is independent of time,[16] some rather special outcomes follow. First, we will be able to express the applied forces as a time-independent 'potential function', $V(\mathbf{r}_i)$, satisfying $\mathbf{F}_i^{appl} = -\partial V/\partial \mathbf{r}_i$ (see Appendix A6.5). Second, we will be able to express the accelerations as $\mathbf{a}_i = \ddot{\mathbf{r}}_i$. Third, as we have already explained, we are choosing to have $\delta\mathbf{r}_i = d\mathbf{r}_i$. Making the appropriate substitutions into (5.7) we obtain:

$$\sum_{i=1}^{N}(-\partial V/\partial \mathbf{r}_i - m_i\ddot{\mathbf{r}}_i)\cdot d\mathbf{r}_i = 0 \tag{5.9}$$

The first term in the bracket, when summed over i, becomes the perfect differential, dV, and, provided that all the masses, m_i, are constant in time, the second term can be re-expressed as follows:

$$\sum_{i=1}^{N}(m_i\ddot{\mathbf{r}}_i)\cdot d\mathbf{r}_i = \sum_{i=1}^{N}\frac{d}{dt}(m_i\dot{\mathbf{r}}_i)\cdot\dot{\mathbf{r}}_i dt = \frac{d}{dt}\left(\frac{1}{2}\sum_{i=1}^{N}(m_i\dot{\mathbf{r}}_i^2)\right)dt \tag{5.10}$$

Does this remind us of anything? Yes, $\frac{1}{2}\sum(m_i\dot{\mathbf{r}}_i^2)$ is the familiar kinetic energy function, T, and so the right-hand side of (5.10) is $\frac{d}{dt}(T)dt$, which is the perfect differential, dT. So we can now rewrite (5.9) as:

$$dV + dT = d(V + T) = 0 \tag{5.11}$$

Finally, this can be integrated to give

$$T + V = \mathit{constant} = E \tag{5.12}$$

This is a familiar but fundamental result - the sum of the kinetic and potential energies of a mechanical system is a constant. To extract the

[16] We mean no explicit dependence on time; obviously there is still an implicit dependence on time, unless we are talking of 'statics'.

importance of this, let's collect together our starting assumptions: we have a mechanical system made up from particles and forces which can be modelled using rectangular coordinates; the applied forces derive from a time-independent potential function; the kinematical conditions remain constant in time; and the masses are also constant. One may think, well, it's hardly surprising that we have a conservation principle given that we have assumed that everything is independent of time, that is to say, conserved. But this is to miss the point; what we have found is that, granted all the above assumptions, the well-known and fundamental law of the conservation of energy actually *follows* from d'Alembert's Principle.

5.6 Review

D'Alembert, that brilliant but "sinister character",[17] discovered a Principle that causes some physicists to shudder even while, at the same time, it answers to almost all of physics: "All the different principles of mechanics are merely mathematically different formulations of d'Alembert's Principle".[18] In d'Alembert's Principle we have a method of treating a problem in dynamics as if it was a problem in statics (using the Principle of Virtual Work). This is down to d'Alembert's "stroke of genius" - treating reversed massy accelerations as reactive 'inertial' forces. These inertial forces are added to the applied forces, and then the condition for 'equilibrium' is found in the usual way. Well, not quite in the usual Newtonian way: instead of the applied and reactive forces cancelling each other out in a pair-wise fashion, we generally have applied forces acting at just a few mass-points, and a melée of inertial forces, one for every accelerating mass. A further departure from Newtonian Mechanics is that instead of arriving at a balance of forces, we have a balance of *energies* - specifically, virtual work.

The tenets of d'Alembert's Principle are:

(1) The mechanical system is made up out of particles and forces, given in rectangular coordinates,
(2) there are such things as 'constraints', and these are, in essence, forces,
(3) moreover, these constraint forces are *reactive* forces,

[17] Truesdell C, 'The rational mechanics of flexible or elastic bodies, 1638–1788', *Leonhardi Euleri, Opera Omnia,* 2nd series, page 186.

[18] Lanczos, page 92.

(4) massy reversed accelerations are, in essence, forces,
(5) moreover, these massy reversed accelerations, known as 'inertial forces', are *reactive* forces,
(6) at any instant, the total virtual work of the reactive constraint forces is zero, for virtual displacements that are in harmony with the constraints (Postulate A),
(7) at any instant, the total virtual work of the applied forces and the reactive inertial forces is zero, for virtual displacements that are in harmony with the constraints.

The new modelling of physical reality is, at first acquaintance, rather strange, especially the newly introduced 'inertial force' which sounds oxymoronic, allying inertness with activity. However, this inertial force has been experimentally detected in countless scenarios (witness the centrifugal, and Coriolis forces, and the inertial force in Section 5.2) and, more than this, it brings in physical insight and philosophical 'correctness'. First, d'Alembert's Principle shows that, whether inertial or constraint, *all* the *non*-applied forces are 'reactive'. Second, we have seen that d'Alembert's Principle is sensitive to '$\mathbf{F}_i^{appl} + \mathbf{I}_i$' rather than to $\mathbf{F}_i^{appl}$ and $\mathbf{I}_i$ taken separately (equation (5.7)). This means that there is no fundamental way of separating $\mathbf{F}_i^{appl}$ and $\mathbf{I}_i$, and so there is no way of determining that an applied force is absolutely external, or that a frame of reference is absolutely stationary, or that a frame of reference is absolutely accelerating. Considering $\mathbf{I}_i$ itself, we see also that the Principle is sensitive to $m_i\mathbf{a}_i$ rather than to m_i and $\mathbf{a}_i$ taken separately. It is therefore impossible to cleanly excise m_i away from $\mathbf{a}_i$, and this leads to a departure from Newton's passive, inert, conception of mass. For 'fictitious forces' (see Section 5.4), the strength of the force, far from being independent of mass, is actually proportional to it. In the special case of gravity this leads to an identity between the 'gravitational charge', m_{grav}, and the mass, m (see the footnote for definitions[19]). Newton noticed this as a coincidence but for Einstein it was one of the clues that led him to the Equivalence Principle and then on to General Relativity. Also, in the centrifugal force, we see a dependence of the force on the *speed*, and the

[19] The 'gravitational charge' occurs in Newton's Law of Universal Gravitation, $\mathbf{F}_{grav} = Gm_{grav}M_{grav}/\mathbf{r}^2$; the 'inertial mass', m, is what occurs in $\mathbf{F} = m\mathbf{a}$. Usually, whenever we say 'mass' we shall mean the 'inertial mass', m.

spatial distribution of mass, and for Coriolis forces, a dependence on the *speed*, and *direction* of the moving mass - all non-Newtonian features.

While some physicists have imagined that d'Alembert's Principle is included within Newtonian Mechanics, and makes no further physical claims, assertions (4) to (7) are extra postulates which go beyond Newton's Laws of Motion. The wider reach, increased explanatory power, and greater profundity of the Principle is not in doubt, and in modern physics we find that Newton's Third Law doesn't always apply, mass isn't always constant, external conditions can vary with time, and energy is not always conserved (as the system isn't necessarily closed), but it *is* always the case that the total virtual work of the 'dynamical' forces is zero (condition (5.8), Section 5.1). We can even go a step further than d'Alembert, and switch from a consideration of forces to virtual work, and recast the Principle as $\sum \delta\omega^{dyn} = \delta\omega_{total}^{dyn} = 0$. This is the cornerstone of classical, relativistic, and quantum mechanics, and so Lanczos's words, repeated again here, are explained: 'All the different principles of mechanics are merely mathematically different formulations of d'Alembert's Principle" (reference given earlier).

Given all this, we may wonder why d'Alembert's Principle isn't better known. The answer has little to do with the physical content of the Principle, and everything to do with the difficulty in solving it (obtaining the mathematical solutions). This is because of two surprising limitations: (1) the Principle still preserves a relic from Newton - it employs particles, forces, and rectangular coordinates, which leads to reactive accelerations that are 'imperfect differentials',[20] and these are not amenable to mathematical solution; (2) despite our vaunting of it, d'Alembert's Principle, being a special application of the Principle of Virtual Work, still applies only at *one* instant of time - curiously inappropriate for dynamics. An ingenious way of overcoming both limitations in one go is given in the next development, known as Lagrangian Mechanics.

[20] Unlike the applied forces, the inertial forces are *not* in general described by a mathematical function, say, y, and so the inertial accelerations are *not* the 'd-of-y' - see the footnote in Section 5.5, and also Section 3.7.2.

6

Lagrangian Mechanics

Joseph-Louis Lagrange, in his great work, the *Mécanique analytique* (Analytical Mechanics),[1] of 1788, discovered a general method for solving all the problems of mechanics, whether in statics or dynamics.

6.1 Introduction

The Principle of Virtual Work yields the condition for static equilibrium: it applies at one instant and then for all time (in other words, time doesn't come into it). D'Alembert's Principle, being a special case of the Principle of Virtual Work, also applies at just one instant but as we're now in the realm of dynamics the conditions do change with time and so d'Alembert's Principle must be reapplied at the very next instant, and then again at the next instant, and so on and so on. However what we would like is a method that frees us from the need to explicitly re-apply d'Alembert's Principle, and, instead, enables us to mathematically track the motions continuously and over the whole time-interval of the problem. For example, in this more 'holistic' method we could have a provisional solution (a guess) spanning the *whole time* of the given dynamical problem (from t_{start} to t_{end}), and then we could 'vary' our whole-solution (have a guess that was a slightly different whole-solution), and then have some criterion for filtering out all but one of our guesses - the answer. Does this remind us of anything? Yes, it reminds us of minimum-path and isoperimetric problems - such as Dido's question, Johann Bernoulli's quickest-descent curve, and the shape of the catenary (see Chapter 2) - all solved using the method known as the 'calculus of variations' (Sections (3.7.2) and (3.7.3), Chapter 3). But perhaps we could use this same method, the calculus of variations, also in the case of general mechanics problems,

[1] Lagrange J-L, *Mécanique analytique* 1788, translated by Boissonade and Vagliente, Kluwer Academic, 1997.

The Lazy Universe. Jennifer Coopersmith, Oxford University Press (2017).
 DOI 10.1093/acprof:oso/9780198743040.001.0001

as follows: we could embed d'Alembert's Principle in a time integral, and then determine that whole-path through time which makes the integral stationary?

This is, in fact, exactly what we will do but there is one stumbling block that stands in our way: the calculus of variations involves analytical (mathematical) techniques and these can only be used when all the equations are in *functional* form (the f and F in equation 3.7.3-1 must be *functions*). The resolution? It turns out that we shall succeed if we restrict ourselves to just those cases where everything *is* in functional form. Thus, we shall insist that the applied forces arise from a work *function*, U, itself arising from a potential energy *function*, $-V$, (for example, such that $\mathbf{F}_i^{appl}(\mathbf{r}_i) = -\partial V/\partial \mathbf{r}_i$), and we shall also insist that any constraints and kinematical conditions are given in functional form.[2]

You may feel that this is all very well but we'll just end up with a mechanics that only works when it works. However, this turns out to be a worthwhile exercise as 'when it works' seems to encompass most of physics! Better still, 'when it works' yields results that are stupendous, seeming to reward us with more physics, more insight, than ever we put in. So let's run with it and see what happens.

6.2 Hamilton's Principle - The Principle of Least Action

At any one time, t, d'Alembert's Principle equates the total 'dynamical' virtual work, $\delta\bar{\omega}_{total}^{dyn}$, to zero (end of Chapter 5)[3] but, in dynamics, we require to integrate $\delta\bar{\omega}_{total}^{dyn}$ through time between endtimes t_a and t_b, and then find the solution, that is, we require to solve $\int_{t_a}^{t_b} \delta\bar{\omega}_{total}^{dyn} dt = 0$. However this integral is not solvable as it stands, and this is because $\delta\bar{\omega}_{total}^{dyn}$ is an *imperfect* differential (Chapter 3, Section 3.7.2 c). It is imperfect for the reason that it contains the work done by the 'inertial' or 'reactive' accelerations, and these have arisen from the combined

[2] There are a few exceptions, known as non-holonomic conditions, which we'll come to later. Why does the potential energy have a minus sign? This is discussed in Section 6.6.

[3] Notation: 'Dynamical' has been defined at the end of Section 5.1, Chapter 5. We denote the dynamical virtual work of one particle at one time as $\delta\bar{\omega}_i^{dyn}$, and for all particles at one time as $\sum \delta\bar{\omega}_i^{dyn} = \delta\bar{\omega}_{total}^{dyn}$. The bar over the omega reminds us that the differential is imperfect, and 'total' refers to 'all particles', not 'all times'.

influence of *many* sources. (Lanczos uses the Greek term 'polygenic' - arising from many sources - as opposed to 'monogenic' - arising from just one source.) But not all the contributions to $\delta\bar{\omega}_{total}^{dyn}$ are imperfect: if only we could arrange things so that we collect together all the perfect bits that are in functional form, and tease out and isolate the difficult imperfect bits. Here's how.

We consider a system comprising N particles, m_i, with (rectangular) position vectors, $\mathbf{r}_i$. We start by integrating d'Alembert's Principle, (5.7), Chapter 5, from time t_a to time t_b:

$$\int_{t_a}^{t_b} \delta\bar{\omega}_{total}^{dyn}\, dt = \int_{t_a}^{t_b} \sum_{i=1} (\mathbf{F}_i^{appl} - m_i\mathbf{a}_i) \cdot \delta\mathbf{r}_i dt = 0 \tag{6.1}$$

(As $\delta\bar{\omega}_{total}^{dyn}$ at each instant is zero, then the integral of it over time is also zero.) The bracket in (6.1) can be separated into two terms:

$$\int_{t_a}^{t_b} \sum_{i=1} \mathbf{F}_i^{appl} \cdot \delta\mathbf{r}_i dt - \int_{t_a}^{t_b} \sum_{i=1} m_i\mathbf{a}_i \cdot \delta\mathbf{r}_i dt = 0 \tag{6.2}$$

Now, one of our assumptions is that there is a potential function, V, with perfect differential, $\delta V = \sum(\partial V/\partial\mathbf{r}_i)\cdot\delta\mathbf{r}_i = -\sum\mathbf{F}_i^{appl}\cdot\delta\mathbf{r}_i$ where $\mathbf{F}_i^{appl} = -\partial V/\partial\mathbf{r}_i$. The first term of (6.2) is therefore already in functional form:

$$\int_{t_a}^{t_b} \sum_{i=1} \mathbf{F}_i^{appl} \cdot \delta\mathbf{r}_i\, dt \qquad \text{may be written as} \qquad -\int_{t_a}^{t_b} \delta V dt$$

The second term in (6.2) is problematic as it contains the dreaded imperfect inertial accelerations - but we have some 'tricks' up our sleeve. First, we ameliorate matters by re-branding the accelerations as velocities:

$$\mathbf{a}_i \qquad \text{may be written as} \qquad \dot{\mathbf{v}}_i \qquad \text{may be written as} \qquad \tfrac{d}{dt}(\mathbf{v}_i)$$

This re-branding 'trick'[4] is one we'll resort to again and again (see especially Chapter 7), and amazingly it works. (The reason why it works

[4] Such a 'trick' was first used by J F Riccati (1676–1754). (The reference is in "When Least is Best", Nahin, p243.)

is hard to explain but, ultimately, it is due to the fact - as Hamilton intuited - that 'position' and 'momentum' are *the* fundamental variables, and so 'speed' (part of momentum) is a more fundamental and telling parameter than 'acceleration'. We'll return to this in Chapter 7.) Finally, assuming that the m_i are constant in time, we obtain:

$$\int_{t_a}^{t_b} \sum_{i=1} m_i \mathbf{a}_i \cdot \delta \mathbf{r}_i \, dt \quad \text{may be written as} \quad \int_{t_a}^{t_b} \sum_{i=1} \frac{d}{dt}(m_i \mathbf{v}_i) \cdot \delta \mathbf{r}_i \, dt$$

The second 'trick' we use is a mathematical technique known as 'integration by parts':

$$\int_{t_a}^{t_b} \sum_{i=1} \frac{d}{dt}(m_i \mathbf{v}_i) \cdot \delta \mathbf{r}_i \, dt = \int_{t_a}^{t_b} \sum_{i=1} \frac{d}{dt}(m_i \mathbf{v}_i \cdot \delta \mathbf{r}_i) \, dt - \int_{t_a}^{t_b} \sum_{i=1} m_i \mathbf{v}_i \cdot \frac{d}{dt}(\delta \mathbf{r}_i) \, dt \tag{6.3}$$

There is scope for modifying the last term in (6.3) by exploiting the fact that the operations $\frac{d}{dt}$ and δ may be swapped around.[5] This term may then be transformed as follows (taking note of the fact that $\frac{d}{dt}(\mathbf{r}_i)$ is the same as $\mathbf{v}_i$):

$$\int_{t_a}^{t_b} \sum_{i=1} m_i \mathbf{v}_i \cdot \frac{d}{dt}(\delta \mathbf{r}_i) dt = \int_{t_a}^{t_b} \sum_{i=1} m_i \mathbf{v}_i \cdot \delta \frac{d}{dt}(\mathbf{r}_i) dt = \int_{t_a}^{t_b} \sum_{i=1} m_i \mathbf{v}_i \cdot \delta \mathbf{v}_i dt \tag{6.4}$$

Now, δ is an operation which, while it relates to the variational calculus (Sections (3.7.2b) and (3.7.3)), it nevertheless follows exactly the same rules as $\frac{d}{dt}$. So we know that

$$m_i \mathbf{v}_i \cdot \delta \mathbf{v}_i \qquad \text{may be written as} \qquad \delta\left(\tfrac{1}{2} m_i \mathbf{v}_i^2\right)$$

This last expression, $\frac{1}{2} m_i \mathbf{v}_i^2$, rings bells for us - it is the kinetic energy of particle, i, and we will denote it by T_i, as T is a symbol commonly

[5] This is acceptable because the process of variation involves only *infinitesimal* changes to the starting function. (And if $\frac{d}{dt}$ and δ can't be swapped, everything is then so pathological that the methods of analytical mechanics fail anyway.)

used for kinetic energy. (δT_i is then the variation in the kinetic energy of particle i.) Summing over all the masses (that is, over all the is) we obtain the virtual variation in the total kinetic energy, δT.

Returning to equation (6.1), we substitute in all of these new expressions, and so finally arrive at:

$$\int_{t_a}^{t_b} \delta\bar{\omega}_{total}^{dyn} dt = -\int_{t_a}^{t_b} \delta V dt - \int_{t_a}^{t_b} \sum_{i=1} \frac{d}{dt}(m_i \mathbf{v}_i \cdot \delta \mathbf{r}_i) dt + \int_{t_a}^{t_b} \delta T dt$$
$$= 0 \tag{6.5}$$

We have made huge progress. T and V are *functions*, and so we have succeeded in isolating and herding together all the difficult imperfect polygenic responses - they're all in the middle term on the right-hand side of (6.5). Even more fortuitous, this middle term, by this method of 'integration by parts', is a 'total differential' and so all of it becomes a boundary term:

$$\int_{t_a}^{t_b} \sum_{i=1} \frac{d}{dt}(m_i \mathbf{v}_i \cdot \delta \mathbf{r}_i) dt = \left[\sum_{i=1} (m_i \mathbf{v}_i \cdot \delta \mathbf{r}_i) \right]_{t_a}^{t_b} \tag{6.6}$$

This helps - the messy bits are concentrated in just one term of (6.5), and even within this term they are concentrated (exist only) at the boundaries (the end-points of the path). But there's still one more 'trick' up our sleeve. It so happens that at these boundaries there is no variation. (Hamilton insisted on this - the integral in (6.5) was to be minimized for a function that remains fixed at the ends (fixed in both position and time[6]).) But this then means that the $\delta \mathbf{r}_i$s are zero at t_a and zero at t_b, and so the whole boundary term is zero - it just disappears, pouffe! We have had one difficult bundle of imperfect inertial accelerations, but we never have to unwrap it, we can throw the whole bundle away. Equation (6.5) thus simplifies to:

$$\int_{t_a}^{t_b} \delta\bar{\omega}_{total}^{dyn}\, dt = \int_{t_a}^{t_b} \delta(T - V)\, dt = 0 \tag{6.7}$$

[6] This is in contrast to Lagrange who allowed some small slack in the final end-time, just enough to make sure that energy-conservation was satisfied.

Now, in the same way as the order of $\frac{d}{dt}$ and δ may be swapped (see above), the order of $\int$ and δ may be swapped. So, the right-hand side of equation (6.7) finally turns into:

$$\textbf{Hamilton's Principle} \qquad \delta\int_{t_a}^{t_b}(T-V)\,dt = 0 \qquad (6.8)$$

This equation has become known as Hamilton's Principle, after the Irish genius and mathematical physicist, William Rowan Hamilton (1805–65) - see Chapter 2. We shall hear more of Hamilton later (in the next chapter). As the integrand, $(T-V)$, has the dimensions of energy, and as the integral is over time, then the left-hand side of equation (6.8) is a quantity of *action* - it asserts that the variation in the total action is zero, or, in other words, the action is stationary. In fact, we shall find (Section 6.6) that the action is not just stationary, it is usually minimized. So, Hamilton's Principle is thus a **Principle of Least Action** (we have the authority of Feynman to name it thus[7]). We at last have our Principle in the form of an equation - but now we must solve this equation. This is dealt with in the next section.

6.3 The solution of Hamilton's Principle: Lagrange's Equations of Motion

We start by making Hamilton's Principle more compact by writing 'L' as a shorthand for '$T-V$':

Hamilton's Principle

$$\delta\int_{t_a}^{t_b}(T-V)\,dt = 0 \qquad \text{may be written as} \qquad \delta\int_{t_a}^{t_b}L\,dt = 0$$

The letter 'L' was chosen to commemorate Lagrange, who formulated a least action principle which was the precursor of Hamilton's Principle. 'L' is known as 'the Lagrangian'. Now T and V are functions of $\mathbf{r}_j$, $\dot{\mathbf{r}}_j$, and t, and so L is likewise some function of $\mathbf{r}_j$, $\dot{\mathbf{r}}_j$, and t, where j runs over the number of particles in the system, say, N. (A historical note: in Lagrange's earlier version L was not allowed to be a function of t;

[7] Richard P Feynman, The *Feynman Lectures* on Physics, Vol II, Chapter 19.

Hamilton's Principle is more general in this regard.) Thus Hamilton's Principle, with all the arguments shown, becomes:

$$\delta \int_{t_a}^{t_b} L(\mathbf{r}_1, \mathbf{r}_2, \ldots \mathbf{r}_N; \dot{\mathbf{r}}_1, \dot{\mathbf{r}}_2, \ldots \dot{\mathbf{r}}_N; t)\, dt = 0 \qquad (6.9)$$

Let's refresh our memory of the calculus of variations from Section (3.7.3), Chapter 3. There we showed that the stationarity of a definite integral (with respect to infinitesimal variations in the shape of the line), required that:

$$\delta \int_{x_a}^{x_b} F(f_1, f_2, \ldots f_n; \dot{f}_1, \dot{f}_2, \ldots \dot{f}_n; x)\, dx = 0 \qquad (6.10)$$

In that section we claimed that, so long as the f_j and the $\dot{f}_j$ were *functions* of x, and so long as F was a *function* of the f_j, $\dot{f}_j$, and x, then (6.10) could be solved by the Euler-Lagrange Equations.[8] Now equations (6.9) and (6.10) bear more than a passing resemblance to each other. In fact, given that L is some, as yet, unspecified *function* of $\mathbf{r}_j$, $\dot{\mathbf{r}}_j$, and t, and F is some unspecified *function* of f_j, $\dot{f}_j$, and x, then, if the Euler-Lagrange Equations solve (6.10), might they not also solve (6.9), that is, solve Hamilton's Principle? Yes, but there is even one more remarkable step we can take: as L is 'some function' of $\mathbf{r}_j$, $\dot{\mathbf{r}}_j$, and t, so surely it is also 'some other function'[9] of q_i, $\dot{q}_i$, and t, where the q_i s are Lagrange's 'generalized coordinates'.[10] We can thus write Hamilton's Principle, in its most general form, as:

$$\delta \int_{t_a}^{t_b} L(q_1, q_2, \ldots q_n; \dot{q}_1, \dot{q}_2, \ldots \dot{q}_n; t)\, dt = 0 \qquad (6.11)$$

and we can be assured that it will be solved by the Euler-Lagrange equations:

[8] Equations (3.11, Chapter 3). Note: there's also the requirement that the functions be continuous, differentiable and finite, and that the boundary conditions are satisfied.

[9] This functional nature of L is guaranteed as, in addition to our insistence that T and V are functions, we also insist that the *transformation* between $\{\mathbf{r}_j, \dot{\mathbf{r}}_j, t\}$ and $\{q_i, \dot{q}_i, t\}$ must be by functions. Note that the limits of j and of i are not necessarily the same ($j = 1$ to N, whereas $i = 1$ to n).

[10] The generalized coordinates are explained in Sections 3.1 and 4.7.

$$\frac{d}{dt}\left(\frac{\partial L}{\partial \dot{q}_1}\right) - \frac{\partial L}{\partial q_1} = 0$$
$$\frac{d}{dt}\left(\frac{\partial L}{\partial \dot{q}_2}\right) - \frac{\partial L}{\partial q_2} = 0$$
$$\vdots$$
$$\vdots$$
$$\vdots$$
$$\frac{d}{dt}\left(\frac{\partial L}{\partial \dot{q}_n}\right) - \frac{\partial L}{\partial q_n} = 0 \qquad (6.12)$$

(It is understood that all these n equations must be satisfied *simultaneously*.) As the independent variable is now the time, t, then (6.12) are equations of *motion*, and by convention the Euler-Lagrange Equations are then referred to as the 'Lagrange Equations of Motion' or simply as the 'Lagrange Equations'. These famous equations, the solution to Hamilton's Principle, can be written on one line as:

The Lagrange Equations of Motion

$$\frac{d}{dt}\left(\frac{\partial L}{\partial \dot{q}_i}\right) - \frac{\partial L}{\partial q_i} = 0 \qquad i = 1, 2, \ldots, n \qquad (6.13)$$

This is all very impressive, but we are left wondering whether we haven't been blinded by wool: can a mere change of symbols ($F \to L$, $f \to q$, $\dot{f} \to \dot{q}$, and $x \to t$) really mean that we have gone from a useful technique (for solving, for example, Dido's problem, the brachystochrone problem, and the shape of the catenary) to a set of equations promising a solution to *all* of mechanics? The resolution will lie in the fact that, in going from the Euler-Lagrange Equations to the Lagrange Equations of Motion, we have done much more than a mere change of symbols. Before this is explained, we'll first treat a few simple and well-known mechanics problems by Hamilton's Principle and the Lagrange Equations of Motion - in other words, by the method known as 'Lagrangian Mechanics' - so that we can be duly awed by the power of our new weapons. (See the worked examples in Appendix A6.1).

6.4 What is happening physically

From solving problems (Appendix A6.1) we learn so much: that what counts as kinetic or potential energy depends on the frame of reference; that it isn't necessary to calculate the constraint forces (the tension in a cord, the force that constrains a bead to a wire, and so on); that the same method and insights can be applied to very different systems (for example, electrical and mechanical systems); that the kinetic energy can contain constant terms, and terms that are linear in the speeds; that the potential energy can depend on velocity, and on time. But even without solving a single problem, something commands our attention and leaves us awestruck: run your eye over all the boxes shaded grey in Appendix A6.1 and you will be witness to a most remarkable thing - the applicability of just *one set of equations*. The systems are for different scenarios, each with their own forms for T and V, and yet the equations to be solved always have exactly the same form. Also, for any one system, the coordinates may be transformed,[11] or the system may be re-modelled into a different set of coordinates, and there may be time-dependent conditions imposed on these coordinates, and the total energy may not be a constant, and the reference frame may be moving, even accelerating; and yet in all these cases the equations to be solved - the Lagrange Equations - still have the same form, *and* the Lagrangian keeps its same value (say, 0.0561 Joules).[12]

Yet we are left feeling puzzled and circumspect: surely, merely by 'sleight of maths' (all the 'tricks' of Section 6.2, and the symbol-changes of Section 6.3), we can't have arrived at a *universally* valid set of equations? To be able to minimize a definite integral, that is a very pretty skill, mathematically-speaking, but our perplexity has nothing to do with the mathematics, and everything to do with the application of it to physics. The perplexity will be resolved when we understand that some implicit *physical* assumptions have entered in along the way. Yes, we can re-model the system, or transform from one set of coordinates to another, but there are some underlying physical aspects that don't change - the 'degrees of freedom'. Yes, we can introduce some extra mathematical 'conditions' (equations between variables) but again something physical is implicated - the number of these 'degrees of

[11] We are not allowing a change of units, like going from centimetres to inches.

[12] But see the optional reading in Section 6.6.

freedom' is reduced. Yes, imposing certain boundary conditions gets rid of the 'boundary term' at a stroke - but what does this mean physically?

Let's address the last question first. We remember Lanczos's Postulate A - the 'reactive' constraint forces, leading to imperfect accelerations, do no virtual work. Now in the boundary term we again have a collection of imperfect accelerations (this time resulting from 'reactive' inertial forces), and we again find that they contribute nothing to the total virtual work. So it seems as if the boundary conditions are a special kind of constraint condition - and this makes plausible physical sense.

Note that Postulate A doesn't apply to the perfect differential motions - we find that δT does not usually lead to zero variation in action; it is only δT *in tandem* with $-\delta V$ that leads to stationary action. Remarkably, this applies both through time and at each instant of time (between the given end-times). That T and V are up to the task is, in a sense, not too surprising; we earlier on stipulated that they must be functions, and, as such, they impart a functional dependence on time.[13] Combined with the boundary and continuity conditions, and maybe also extra constraint and kinematic conditions, this functional dependence is exactly such as to rule in one actual path, and to weed out all other paths from our enquiries. (Moreover, T and V are still up to the task, whether they are functions of $\{q_i, \dot{q}_i, t\}$ or $\{\mathbf{r}_j, \dot{\mathbf{r}}_j, t\}$.)

Another mathematical stipulation, introduced rather casually (just before Hamilton's Principle, equation (6.8), Section 6.2), was that the order of $\int$ and δ may be swapped. This seemingly innocent swapping operation has *physical* implications: we go from virtual *displacements* at one instant, to a varied *whole-path* over a prescribed time-interval. Now, one may at first think that this extra whole-path requirement is unnecessary - surely, as the total virtual work is already zero at each and every instant,[14] then it is guaranteed to be zero over the whole path in time? Yes, but while a minimized[15] path does require that every sub-interval must also be minimal, this doesn't mean that *any* collection of minimal sub-intervals will be just the right collection to yield the required whole path. In other words, the vanishing of the virtual work

[13] This dependence is implicit - through the time-dependence of the motions - but could also be explicit; that is, there is nothing we have said that prevents T or V depending *directly* on t. (That is, we could have $T = T(t)$, or $V = V(t)$.)

[14] (by d'Alembert's Principle, which is our starting principle)

[15] The same argument applies to a stationary path, but it is more intuitively obvious when considering a minimum path.

at every instant is a necessary but not a sufficient condition to generate the correct final path. This gives us some insight into an earlier question: at the end of Chapter 3 we wondered why the varied integral had $\dot{q}$ as one of its arguments - surely a dependence on q is enough, as $\dot{q}$ is merely consequential upon q? But now we begin to understand: in order for a collection of infinitesimal path segments to add up to the correct whole path, then each segment must join on smoothly and continuously with the next segment, and so on. This can only be guaranteed if the $\dot{q}$-values are equal at the join. Or, in other words, the $\dot{q}$-values are like an infinity of 'interior boundary conditions', existing all the way along the path. (The boundary conditions at the very ends are determined in another way - by the limits on the integral.)

What of the scalar energy functions, T and V? As mentioned earlier, we find that the total virtual work of variations in T cancels out with the total virtual work of variations in V, both instant by instant, and over the whole time of the given problem. But, in fact, we have no need to keep harking back to the form of energy known as *work*: we have progressed to a more general vision - variations in energy, pure and simple (δT and δV). Now, taken individually (that is, for each i), the T_i will not necessarily be the same as they were when modelling the system as point-particles and rectangular coordinates (compare T in examples (1) and (2), Appendix A6.1). Also, V may change as we go from one coordinate system to another. However the total Lagrangian, L, (defined as $(T - V)$), *will* be the same (have the same actual value, say 10 Joules). We say that L is *invariant*. This means that for any one system, howsoever that system is viewed or modelled, the L-value will remain unchanged.[16] We'll comment again on this wondrous outcome in Section 6.6, and discuss the forms of T and V in Section 6.5.

Let's summarize some of the physical assumptions implicit in Lagrangian Mechanics:

1) the system can be modelled as 'generalized particles', q_i, having individual 'perfect motions' with an, as yet unknown, *functional* dependence on time,
2) these individual 'perfect motions' imply an amount of 'work' which makes up one all-embracing scalar *function* known as T. Moreover, T is an *energy* function (known as the kinetic energy).

[16] However, see the optional reading in Section 6.6.

The functional form of T depends on the modelling (the choice of $\{q_i\}$), and is known in advance,

3) the configurations and interactions[17] of the $\{q_i\}$, in other words the *whole-system* aspects, make up one all-embracing scalar *function*, V, which, of its nature, depends on the system, and is known in advance. Moreover, V is an *energy* function (known as the potential energy).
4) there may, as well, be extra constraints or conditions on or between the generalized particles, and these will manifest themselves in the form of known *functional* relationships between the q_i (the functions may also depend on $\dot{q}_i$ or t),
5) the melée of imperfect reactive motions due to constraints 'cancel each other out' (their total virtual work is zero) at each instant, t,
6) the melée of imperfect reactive motions due to boundary conditions also leads to zero virtual work at each instant, t,
7) the virtual variations, δT and -δV, when these are considered *together*, bring the variation in total action to zero, at each instant, and over the whole time interval of the given problem,
8) the only functions ('equations of motion') for the $\{q_i\}$ that can survive assumptions 1) to 7), along with the requirements of continuity, and boundary conditions (and also differentiability, and being finite), are the ones describing the actual motions.

That Nature does, in fact, conform to these requirements is borne out by the success of Lagrangian Mechanics and of the Principle of Least Action in physics.

6.5 The functions T and V

The generalized coordinates for a given system can usually be arrived at after thoughtful inspection of that system and seeing the 'motions' of which it is capable. But what of the functions T and V? We have merely said that they are supplied beforehand. Now we will examine what form they must take.

[17] Note that these configurations and interactions can be with respect to *time*, *speed*, or *direction*, as well as with respect to position.

6.5.1 *The form of T*

For a single free particle (no forces acting) of mass, m, and velocity, $\mathbf{v}$ (in rectangular coordinates), the kinetic energy has the familiar form:

$$T = \frac{1}{2}mv^2 \qquad \text{kinetic energy, 'quadratic form'} \qquad (6.14)$$

Why does T have this 'quadratic form', is it essential? (The adjective 'quadratic' is to remind us that the speed is *squared*.) The physicists Landau and Lifshitz[18] argue that, due to the homogeneity of space and time, the kinetic energy for a free particle cannot depend on position, $\mathbf{r}$, or on time, t. Therefore (they argue), T must depend only on the velocity, $\mathbf{v}$. Furthermore, the isotropy of space requires that no special directions are picked out, and so T must depend only on the magnitude of the velocity, that is, on v. This leads Landau and Lifshitz, finally, to conclude that T must depend only on the speed squared, $T = T(v^2)$.

But there are questions to be asked of Landau and Lifshitz's 'derivation'. First, as well as no absolute position or time, we know that there is also no absolute velocity (it is impossible to tell, by any experiment within a given reference frame, at what uniform speed or direction the reference frame is travelling) - so how can there be a dependence on $\mathbf{v}$ or v, squared or not? Second, why should a magnitude - the speed - only be arrived at by squaring? Why not, say, squaring followed by taking-the-square-root? In fact, it can be shown that a *squared* dependence on speed *is* required. This can be shown in two ways:

(i) A physics argument: The Principle of Relativity must always be upheld, and this can only be guaranteed by having a dependence on v^2. Two proofs of this (due to Maimon,[19] and due to Ehlers et al[20]) are given in Appendix A6.2,

(ii) A mathematical argument: the only way in which to arrive at a scalar invariant quantity from the vector quantity, $\mathbf{v}$, is to take the 'scalar product', $\mathbf{v}\cdot\mathbf{v}$, and this yields v^2.

[18] Landau L, and Lifshitz E, *Course in Theoretical Physics*, vol 1, *Mechanics*, Pergamon Press, Oxford, 1960.

[19] Maimon, Ron: former contributor to Physics Forum on the web; Maimon is an independent researcher in physics.

[20] Ehlers J, Rindler W, and Penrose R, 'Energy conservation as the basis of relativistic mechanics II', American Journal of Physics, 33(12) 1965, pages 995–7.

Upon reflection, we find that both arguments have a common feature: to obtain an absolute or invariant quantity we must always consider one quantity *relative* to another. So, in (i) we find that both proofs require a collision between *two* particles; and in (ii) we note that the scalar product, **a**·**b**, presumes *two* vectors. Even in the case of identical vectors, **v**·**v**, then one **v** is the 'projected' vector, and the other **v** is the 'projectee' vector; in other words, the 'dimensionality' of the problem is *dual*.[21] (In differential geometry, we say that each **v** has come from a different 'vector space'.)

However, a solution can sometimes lead to a new problem: what form shall we give to the kinetic energy of just *one* isolated particle? The resolution is that we are then compelled to *define* the kinetic energy as having the quadratic form (6.14). This leads to consistency in the theoretical modelling. (On the other hand, we could bow out altogether on the grounds that 'one isolated particle' is a pathological scenario - we can never check up on it experimentally because as soon as we bring in measuring equipment then the particle is no longer isolated. Once again, the author's maxim is apt: "[pathologically] simple cases make bad physical intuition.")

We are not done yet; in (6.14), there is still the question of mass, m. How did we arrive at 'T is proportional to m'? To answer, we follow Maimon and use the empirical observations that mass is additive (the mass of many particles is the sum of their individual masses), and that for a particle of, say, double the mass, then its kinetic energy is doubled. These empirical findings mean that not only is T a function of v^2 but mass *must* occur as the coefficient of this function of v^2. So, we end up asserting that: T='massy inertial factor'×'some function of v^2', where the 'massy inertial factor' always includes m.[22] Finally, when we are modelling the situation using generalized coordinates, then the standard form is: T='massy inertial factor'×'some function of $\dot{q}^2$'. (We'll come back shortly to the question of the factor, $\frac{1}{2}$.)

We are still not done; look again at definition (6.14) and ask: How big is the mass, m? How big is the speed, v? For masses that are tiny, or huge, or speeds that are high, then (6.14) needs modification, and we are in the

[21] The word 'dimensionality' will be explained near the beginning of Chapter 7.

[22] There is still the question of how the 'massy inertial factor' is split between the two dual vectors: we could have 1 and $\frac{1}{2}m$, or $\sqrt{m/2}$ and $\sqrt{m/2}$, and so on. The resolution will come with Hamilton's p and q coordinates (Chapter 7), and it will be shown that all the massy-dependence goes to just one of the dual vectors.

realms of quantum mechanics, Einstein's Theory of General Relativity, and Einstein's Theory of Special Relativity, respectively. However, a common feature of all great theories is that their wisdom extends beyond their conventional range of applicability. This is especially true of Special Relativity, which has not merely extended Newtonian Mechanics into a high-speed regime but has radically changed our understanding of all of physics, whatever the speed. (It was from Special Relativity that Einstein arrived at his most important single discovery, $E = mc^2$ - see below.) Of particular relevance in the present discussion: in Newtonian Mechanics there is no absolute speed, but in Special Relativity there *is* an absolute speed - the speed of light, c. However, in all the formulae of Special Relativity v always occurs as the ratio, v/c, - so speed has, in effect, been converted into a 'relative quantity'. Should we be concerned that in the standard form for T, (6.14), speed occurs on its own (that is, without c)? No, we are rescued in a different way. The method of Lagrangian Mechanics (and all Variational Mechanics) concerns a *system*, and we find that the speed is always relative to that system. In fact, in the Variational Mechanics, all the coordinates are relative to the end-points of the integral in Hamilton's Principle, and so *all the coordinates are relative to the system*. This answers our earlier question over Landau and Lifshitz allowing v into the formulation, and is yet another reason why Lagrangian Mechanics rather than Newtonian Mechanics survives the transition to Einstein's Relativistic Mechanics. (By the way, the factor, $\frac{1}{2}$, must be included in order to ensure that the Special Relativistic kinetic energy reduces to the usual formulation of (6.14) in the limit of 'low' speed.)

Saving the most important till last, we come to Einstein's most famous discovery, the best-known equation[23] in physics, $E = mc^2$. Einstein, himself, regarded this discovery - of the equivalence between mass and energy - as the single most important result to emerge from Special Relativity.[24] One consequence of this equivalence is that even if the particle slows to a standstill, and its kinetic energy is then zero, the particle still has mass, and so there is still some energy 'left over'. Now, in Newtonian Mechanics, the zero-point of kinetic energy is arbitrary,

[23] We have left this iconic equation in its usual form, but, in fact, it depends on the notation for mass: if we use 'm' to mean 'rest mass', then the equation should be written $E = m\gamma c^2$, where $\gamma = 1/\sqrt{(1 - v^2/c^2)}$.

[24] Einstein A, in Pais A, *Subtle is the Lord: the Science and the Life of Albert Einstein*, Oxford University Press, 2005.

and is generally set to zero. However, Einstein found that the 'left over' energy, sometimes called the 'rest energy', is not arbitrary - it is determined, and specific to the given particle.[25] It also happens to be enormous (1 gram of mass has a 'rest-energy' of some 90 million million Joules). Why are we usually so (blissfully) unaware of this huge 'rest energy'? It is because, contained within mass, it happens to be an exceptionally stable way of storing energy. One further question on this: as the 'rest energy' is evidently a store of energy, should we not consider it as V instead of lumping it together with T? We'll say more about this as we go along, and end up with a speculation in the concluding section of the chapter.

Einstein's famous equation applies not only to kinetic energy but implies that *all* energy is 'massy'. This mass-energy equivalence is anticipated in Lagrange's and later Hamilton's Mechanics, which treats kinetic energy (rather than acceleration) as the true measure of inertia (see the next section, 6.6). It is therefore not surprising that Lagrangian Mechanics is well placed to accommodate Einstein's Relativity Theory (it is only necessary to find the appropriate form for L) whereas Newton's Mechanics must be totally replaced. One extra thing to say is that it is an empirical and not fully understood fact that mass, and hence also kinetic energy, are always *positive* quantities.[26]

The standard form for T applies to free[27] particles. When considering more complicated scenarios, for example with interacting particles, or accelerating reference frames, then T *can* have terms just containing q, or unsquared-$\dot{q}$, and so on (see Appendix 6.1, problem (10)). These occurrences do not contradict the 'quadratic' form of (6.14), as there we had a free particle and so space really was featureless (homogeneous); but when we consider interacting particles, and external conditions, then the space of the problem is no longer homogeneous, and so a dependence on $q, \dot{q}, \ddot{q}, \ldots$ can occur.[28]

[25] Of course, it is determined from the mass, in accordance with Einstein's famous equation - as we are talking of a particle, with no internal parts, then we can forget about internal energy states.

[26] W Rindler, *Special Relativity*, Oliver and Boyd, 1966, p 87.

[27] Colliding particles may be considered free at all times except for the instant of collision; also, a uniformly rotating finite-sized rigid body may be considered as free.

[28] (For terms linear in $\dot{q}$, you may be wondering about a contradiction with our earlier footnote that the system relates to a *dual* vector space. However, these linear terms occur when some constraint condition or potential energy term also contains a linear dependence on speed - and so, overall, the duality is maintained.)

6.5.2 *The form of V*

There is a standard form for T, (6.14), but none for V - it must be formulated afresh for each new system. This has led some physicists to speculate that V is in some sense less fundamental than T - more on this later (end of Sections 6.6 and 6.8). Now V arises from applied forces, and from 'particle'-interactions, and as these depend on configuration - the positions of external features, the positions of the 'particles' with respect to one another - then V is explicitly a function of these position coordinates, $V = V(q_1, q_2, \ldots q_n)$. T has already hived off the dependence on the motions, $\dot{q}_i$, (T is, after all, kinetic) and so we generally do not have V depending explicitly on $\dot{q}_1, \dot{q}_2, \ldots$ (but see below). Finally, there is nothing we have said, at any stage in our development of Lagrangian Mechanics, that prevents L, and hence V, depending explicitly on time. So, altogether, our simplest form for V has it depending just on q, and t.

You may be wondering why we are allowing an explicit dependence on q and on t for V but not for T-in-standard-form (in the latter case we (or rather, Landau and Lifshitz) invoked the homogeneity of space and time to veto just such a dependence on q and on t). The explanation is that V, of its nature, describes *whole-system* attributes, and it can therefore depend on whole-system 'configuration' coordinates, whereas T-in-standard-form, of its nature, depends on *individual* or system-free attributes. Put differently, as far as V is concerned, q and t are not absolute but are defined with respect to the given system: the q describe the positions of particles *relative* to each other or to some external constraint, and t is *relative* to the prescribed end-times of the path, t_a and t_b.

Does an explicit time-dependence in V (and, perhaps, in constraint or kinematic conditions) mean that energy conservation is not satisfied? Yes, in Lagrangian Mechanics it is *not* required that energy be conserved.[29] Obviously, in the whole world, energy is still conserved, but an explicit time-dependence in V, or in the external conditions, allows for energy to come into the system from outside. For example: in the 'variable energy cyclotron' the magnetic field is steadily increased to ensure that an accelerating charged particle will stay in the same orbit, but the energy for the magnets must be fed into the system from the outside; in 'Ehrenfest's pendulum', the length of the pendulum is slowly decreased

[29] In Lagrange's own version of Lagrangian Mechanics energy conservation *is* required, but in Hamilton's Principle energy conservation is a special case.

- but hauling in the pendulum takes energy; a spinning turntable introduces centrifugal, Coriolis, and other effects - but the energy to spin the table has been externally supplied; and so on. We'll have more to say on the conservation of energy in Section 6.7.

So much for V in standard form. Now, earlier on, we said that T was responsible for the dependence on speeds, but in fact there is nothing that prevents V depending explicitly on the speed coordinates, $\dot{q}_i$, and there are some famous and important cases. One concerns the motion of a charged particle moving under the influences of electric fields, **E**, and magnetic fields, **B**. The force (known as the Lorentz force) is given by:

$$\mathbf{F} = q[\mathbf{E} + (\mathbf{v} \times \mathbf{B})] \tag{6.15}$$

(This is in units where $c = 1$. The q here refers to electric charge and not to a generalized coordinate). This implies a potential energy given by:

$$V = q\phi - q\mathbf{A} \cdot \mathbf{v} \tag{6.16}$$

where ϕ is some scalar potential and **A** is a vector potential.[30] Note that, in this example, V depends on the velocity, **v**, of the particle (on the direction of motion as well as the speed). Just imagine the consternation when these effects were first discovered - when Oersted, then Faraday, and others, discovered that magnetized compass needles, iron filings, electric currents, and isolated charges could travel along *curved* lines, sometimes in *closed loops* neither starting or ending on a 'source', and (for a pure magnetic field) with the velocity, the field, and the force all *at right angles* to one another. Newton's Mechanics is contravened, but Lagrangian Mechanics carries on as usual (it is only necessary to use Hamilton's Principle with a Lagrangian given by $L = T - q\phi + q\mathbf{A} \cdot \mathbf{v}$).

Another example of velocity-dependent effects are Coriolis forces - those forces partly responsible for wind directions, and ocean currents. For a particle of mass, m, moving at speed, v, with respect to a rotating platform (with constant spin-rate, ω) then the Coriolis force acts *sideways on* to the particle's motion, and is stronger the greater the particle's mass, or speed. In symbols, the Coriolis force has magnitude $2m\omega v_{radial}$, or $2m\omega v_{tangent}$, and is tangential when the velocity is radial, and radial when the the velocity is tangential. (Also, the Coriolis force doesn't depend on the radius, and is distinct from the centrifugal force, another kind of velocity-dependent force.)

30 The reader can find out more from an introductory text on electrodynamics.

In most cases of velocity-dependent potentials the particle's motion violates Newton's Third Law of 'action and reaction', and its momentum is not conserved. However the *generalized* or canonical momentum (Section 6.8, and Chapter 7) *is* conserved, for example, for a moving charged particle,[31] the particle's momentum is not conserved on its own, and the momentum of the electromagnetic field must make up the balance.

6.6 The form of L

The energy function, L, was introduced rather casually in our 'derivation' of Hamilton's Principle (see the beginning of Section 6.3). L, the 'Lagrangian', is the integrand in Hamilton's Principle, and is of paramount importance in mechanics - but where does it come from, and why does it have the form '$(T - V)$' when we are so much more familiar with '$(T + V)$'? We turn to Lanczos,[32] as an explanation of the physical origins of L is given there and nowhere else.

At any instant, T is a function which describes the motions of all the particles,[33] and these motions are a consequence of the 'marching orders' dictated by V. At the very next instant the particles will adopt new positions,[34] and so the configuration has changed, and so the magnitude of V is different. This new V will influence the motion of the particles afresh, and so the T at the next instant will be different. This new T will again lead to the particles adopting new positions, and this will again imply a V with a different magnitude, and so on and so on. We therefore have an interplay between T and V, one which continues, instant by instant, from the start- to the end-time of the problem. Now, as T involves masses-in-motion, it can be seen as an *inertial response* to V (and this is consonant with Einstein's mass-energy equivalence, coming over a hundred years later). This is reminiscent of the inertial response, $m\mathbf{a}$, to Newton's force, $\mathbf{F}$. However, whereas in Newtonian Mechanics we treat one particle in isolation, and end up with an equality (between $m\mathbf{a}$ and $\mathbf{F}$), in Lagrangian Mechanics we have a whole system, and end up with a balancing process between two scalar energy functions. How

[31] (for a ϕ and $\mathbf{A}$ which are independent of position).

[32] Lanczos, bottom of pages 21 and 27, also pages 118–9.

[33] 'particles' can be 'generalized particles'.

[34] This includes not just new positions in space, but different relative speeds, orientations, and so on.

shall we choose between these postulates - Newton's Second Law, and Hamilton's Principle? It is not simply a question of switching from Newtonian to Lagrangian Mechanics when going from simple to complicated scenarios, but rather that Lagrangian Mechanics supersedes and totally replaces Newtonian Mechanics. This is because, despite what we have been coached to believe, it is kinetic energy which is the true measure of 'inertia', and kinetic energy rather than force which is the thing that is truly primitive and irreducible:

$$m\mathbf{a} \qquad = \qquad \mathbf{F}$$

is superseded by

$$T \qquad \text{'balances'} \qquad V$$

(where 'balances' refers to a whole process, the application of a principle, and not merely to 'equality'.)

Finally, why does L have the iconic form $L = T - V$ and no other? It is because, while T and V must be in balance, they must also act *in opposition* to each other. If each could reinforce the other then this could lead to a runaway growth in energy - a very unphysical outcome. But if they act in opposition, or, in other words, if we must find the difference between them, then this can lead to a stable value for energy - a physically plausible outcome. Thus it is the difference between T and V that is important, and therefore V must carry a negative sign. According to Lanczos:

> "the excess of kinetic energy over potential energy is the most fundamental quantity in . . . mechanical problems."[35]

So, T opposes V rather than T and V reinforcing each other - but does this mean that Hamilton's Principle will always lead to *least* action? (We require the action to be stationary, but this still leaves open the question of whether it is a maximum, a minimum, a saddle point, or a plateau.) The answer has been given in an interesting paper by Gray and Taylor.[36] These authors show that we can have a true minimum (a pure minimum), and we can have a saddle point (a minimum and a

[35] Lanczos, page 113.

[36] C G Gray and E F Taylor, When action is not least, AJP 75 (2007) pages 434–58.

maximum together), but it is never the case that there is a true maximum in the action. The argument is made intuitive in the following way. Consider the time-path of a particle in 2-D space using (x, y) coordinates (the one scenario where everyday space and configuration space are the same). Can we vary the path and make it shorter? Usually, unless the path already has the minimum length. Can we vary the path and make it longer? Always, just add more wiggles. Thus there is a difference between a minimum and a maximum: only the former can be guaranteed to be unique. The argument also seems physically intuitive in another way, on the grounds of economy - that is to say, it is more *economical* in action to have it minimized rather than maximized. As supporting evidence, we have the finding that for stable equilibrium $(T = 0)$ V is at a minimum - a well-known result in statics.

Optional

Even when in iconic form, we note that L cannot be defined uniquely, for two reasons. First, the condition of stationarity (the equating to zero of the action integral in Hamilton's Principle) will be unaffected if the whole integral, $\delta\int Ldt$, is multiplied by an arbitrary constant, '*cons*'. Second, the condition of stationarity will likewise be unaffected if the integrand, L, is changed to, $L + df/dt$. This doesn't invalidate what we said earlier about L being an invariant - for any given system, the Lagrangian is not defined uniquely, but whatever choice we plump for then this L *is* invariant (with respect to a different choice of $\{q_i\}$, or to coordinate transformations - excluding re-scaling). This non-uniqueness of L is perhaps the reason why its invariance isn't vaunted as much as the invariance of the Lagrange Equations: these Equations are invariant come what may.[37]

Another comment concerns the sign of L. While T is always a positive quantity (m is positive, and the speed is real) we cannot go on to say that L is always positive, first because V has no absolute sign - it can be set to be positive or negative - and second because the multiplicative constant, *cons*, can be positive or negative.

One intriguing observation is that the condition of stationarity is sensitive to L, that is, to the *whole* of $(T - V)$, rather than to T or V taken separately. Does this imply that there is some blurring of the

[37] (Note that when we have non-holonomic conditions or polygenic forces, such as 'friction', the effect of these non-holonomic influences is to introduce a 'right-hand side' to the Lagrange Equations, see Appendix A6.5, and Lanczos, pages 146–7.)

distinction between what counts as kinetic energy and what counts as potential energy? The answer is yes. For example, in the case of a bead on a wire rotating with constant angular speed, ω (Problem (10) in Appendix A6.1), we have $T = \frac{1}{2}m\dot{r}^2 + \frac{1}{2}mr^2\omega^2$ and $V = 0$, but we could just as well consider it as $T = \frac{1}{2}m\dot{r}^2$ and $V = -\frac{1}{2}mr^2\omega^2$. It makes no difference whatsoever to L, but in the second case we have lost some motional energy of the bead and gained a 'centrifugal' potential that didn't exist beforehand. Such a shift in the assignment of energies can also occur as a result of a change in coordinates (a change in viewpoint), say, from a reference frame sited at the centre of rotation to one sited on the bead itself (see Sections 6.8 and 6.9).

This blurring between T and V is an example of how, in Lagrangian Mechanics, there is no hard and fast distinction made between a whole system and the individual components of that system. For example, consider the motion of a planet in its orbit around the Sun. The planet is a very small body by comparison with the dimensions of its orbit, and with the size of the Sun. In the Newtonian analysis the planet is a 'particle' and doesn't affect its 'surroundings' but in Einstein's theory of General Relativity the planet does distort spacetime to a tiny extent, and this leads to a subtle change in its orbit - its precession advances. (In fact, it was just this tiny effect which led to one of the confirmations of General Relativity - the extra precession of the perihelion of Mercury.) Consider, also, the case of an electron moving parallel to a wire in which a current is flowing;[38] the moving electron feels the magnetic field and is attracted towards the wire. But what happens if the system is viewed from a reference frame which moves such as to cancel out the motion of the exterior electron? The now-stationary electron feels no magnetic force, but, according to the Principle of Relativity, it must still be attracted to the wire. What happens is that the now-moving wire suffers a Special Relativistic effect known as 'length contraction'; the relative density of protons is thereby minutely increased, and so the wire acquires a net positive charge, and attracts the exterior electron, as before. The same overall outcome occurs in both reference frames - and this accords with the Principle of Relativity - but there has has been some subtle switching between 'field energy' and kinetic energy, that is,

[38] The example is taken from Feynman's *Lectures on Physics*, Vol II, 13–7. Note how a relativistic effect occurs ('length contraction') but the speed of the wire is tiny, around 0.001 m/s.

between V and T. The method of Lagrangian Mechanics can be adapted for use in electromagnetism and in gravitation just because it has this whole-system outlook.

Despite the ambiguity between T and V, the former has a fundamental form, equation (6.14), and is always positive, whereas the latter is different for different systems, and has no universal form. This led some nineteenth-century physicists to think that T was in some sense more fundamental than V. Most notably, the great nineteenth-century physicist, James Clerk Maxwell (1831–79), wrote with regards to T,

> "... we are unable to conceive that any possible addition to our knowledge could explain the energy of motion [T] or give us a more perfect knowledge of it than we have already"[39]

> whilst with regards to V,

> "... the progress of science is continually opening up new views of the forms and relations of different kinds of potential energy."[40]

Also, Heinrich Hertz (1857–94), (the first to detect radio waves), went so far as to suggest that there was, at an elemental level, only kinetic energy - in the form of microscopic motions - and that these hidden motions accounted for the effects known as V. (We make brief mention of this again at the end of Section (6.8).)

6.7 Noether's Theorem, and the definition of energy

Let's bypass T and V and consider the relationship of L directly on the generalized coordinates, $L = L(q_i(t), \dot{q}_i(t); t)$.[41] We will now consider those cases in which V, and any external conditions, do not depend explicitly on time. Then, while we still have $q_i = q_i(t)$ and $\dot{q}_i = \dot{q}_i(t)$,[42] nevertheless t will not occur explicitly in V or in the external conditions, and so t will also not occur in the argument of the function for L. Therefore we will have the special time-independent form, $L = L(q_i(t), \dot{q}_i(t))$.

39 Maxwell J-C, *The Theory of Heat,* 1871, Dover Publications, (2001), p 301.

40 *The Theory of Heat,* as above, p 302.

41 This is an abbreviated way of writing $L = L(q_1, q_2, \ldots q_n; \dot{q}_1, \dot{q}_2, \ldots \dot{q}_n; t)$.

42 (These are, in fact, the very 'equations of motion' that we wish to solve in mechanics.)

We postulate that time is homogeneous - there are no absolute markers in it, and no times are more special than other times as regards the applicability of our laws of physics.[43] Pertinent to our present discussion is the law of physics known as Hamilton's Principle, and we find that as time is homogeneous then it should not matter what the absolute end-times of the action integral are. So, we could displace the whole action integral through a small constant time interval, ϵ, and this displacement should make no difference whatsoever to the outcome:

$$\delta \int_{t_a}^{t_b} L(q_i(t), \dot{q}_i(t))\, dt = \delta \int_{t_a+\epsilon}^{t_b+\epsilon} L(q_i(t), \dot{q}_i(t))\, dt = 0 \qquad (6.17)$$

By definition, shifting the limits of integration through $+\epsilon$ is equivalent to shifting ('transforming') the time coordinate through $-\epsilon$. Thus we arrive at:

$$\begin{aligned} \delta \int_{t_a+\epsilon}^{t_b+\epsilon} L(q_i(t), \dot{q}_i(t))\, dt &= \delta \int_{t_a}^{t_b} L(q_i(t-\epsilon), \dot{q}_i(t-\epsilon))\, d(t-\epsilon) \\ &= \delta \int_{t_a}^{t_b} L(q_i(t-\epsilon), \dot{q}_i(t-\epsilon))\, dt = 0 \qquad (6.18) \end{aligned}$$

We could just as well say that we have a new time coordinate, say, $\tau = t - \epsilon$, and it hasn't made any difference to anything. So far so unexceptionable - but it turns out that this is only the beginning.

A hundred years ago, the mathematician Emmy Noether (1882–1935) obtained an outstanding theorem with repercussions across the whole of field theory. Her theorem, in words, is that if the action integral has no explicit dependence on a given coordinate, q_i, or t, then it will be invariant with respect to certain infinitesimal transformations of this coordinate, and the system will then exhibit a symmetry or conservation law concerning this coordinate. A specific symmetry will arise for *any* 'absent' coordinate but here we consider the time coordinate.

We may begin to suspect that these *infinitesimal transformations* are the same as *a variation* (the δ-process, see Chapter 3, Section 3.7), but one aspect that rings a minatory warning is the fact that Noether has included t as one of the coordinates that may be transformed, whereas

[43] (Pretend we don't know about the Big Bang.)

from Section 3.7 we remember that variation of the independent coordinate, t, is *not* allowed. More alarming still, Noether does not restrict the transformation of t to a fixed translation but allows ϵ to be a *function*[44] of time - so we could have $\tau = t - \epsilon(t)$. It is now not at all obvious that our translation in time will make no difference and maintain the homogeneity of time (we are, in effect, allowing time to become bunched up or stretched out between the end-times of the integral).

It turns out, however, that even in the case of such inconstant translations invariance of the action principle can be assured. There is one crucial thing that must be taken into account in this more general case: not only does L change, but dt changes also, and it is no longer sufficient for L on its own to be invariant but rather it is the whole product, Ldt, which must be invariant. We can follow through the mathematics (Appendix A6.3) or jump straight to the result: for a system with time-independent external conditions, the action integral is invariant with respect to an infinitesimal translation in time, $\epsilon(t)$, *provided* that the following conservation rule applies:

$$\left(\sum_{i}^{n} \frac{\partial L}{\partial \dot{q}_i} \dot{q}_i\right) - L = \text{constant} \tag{6.19}$$

What is this mysterious left-hand side that is equal to a constant, in other words, it stays the same for all times? (As neither t_a nor t_b appear in (6.19) then the conservation rule is independent of these specific end-times.) Let's begin to answer this by choosing our simplest time-independent case-study: $T = \sum \frac{1}{2} m_i \dot{q}_i^2$, $V = V(q_i)$, and, by definition, $L = T - V$. Then $\partial L / \partial \dot{q}_i = m_i \dot{q}_i$ for all i, and so the left-hand side of (6.19) becomes:

$$\left(\sum_{i}^{n} m_i (\dot{q}_i)^2\right) - L = 2T - L = T + V \tag{6.20}$$

The constant on the right-hand side of equation (6.19) can then be identified with the right-hand side of equation (6.20), $T + V$, which in turn can be identified with E, the total energy. In other words, we have found that *the total energy is a constant*.

So, we have assumed time-independent conditions and out pops the result that the total energy, E, is conserved - the well-known 'law

[44] (albeit a function that is infinitesimal, continuous, and differentiable)

of the conservation of energy'. The emergence of a conservation law may appear unsurprising given that V and all the external conditions were taken to be time-independent, that is to say, 'conservative'; but remember that we have not brought in energy conservation as an extra condition but have *deduced* it, and only by asserting the validity of Hamilton's Principle, and that time is homogeneous.[45] What will emerge later (in the work of Hamilton, Chapter 7) is that $[\sum(\partial L/\partial \dot{q}_i)\dot{q}_i - L]$ is a more general form for the total energy, and moreover it is not necessary that the total energy satisfies $E = T + V$. Indeed, there can arise mechanical systems which have terms which are linear rather than quadratic in $\dot{q}$, or where T is more complicated than $\frac{1}{2}m\dot{q}^2$. Such systems, if they are independent of time, still satisfy an energy conservation law but it takes the more general form of equation (6.19), which holds for any time-independent Lagrangian.

6.8 External conditions

Every now and again in mathematics, physics, or technology, somebody comes up with a new idea, technique, or gizmo, that is simple to state but is sheer genius. For example, we have Stevin's 'Wreath of Spheres' (Chapter 2), Torricelli's barometer, Pascal's 'triangle', the Newcomen engine, Watt's 'parallel motion', and so on. Lagrange's method for dealing with the 'condition equations' - the 'method of Lagrange multipliers' - is one such a gem. We deal with it in this section.

'Conditions' are *functional* relations, $f(q_1, q_2, q_3, \ldots)$, between the generalized coordinates - examples include 'constraint conditions' (a block slides on a given surface without friction; a cord maintains a certain length; a body is rigid; and so on), and 'kinematic conditions' (a block slides down an inclined plane resting on a trolley, and meanwhile the whole trolley moves uniformly along tracks; a bicycle wheel spins about the axle at a constant rate and in a plane at right angles to the spin-axis; and so on). So far in this chapter we have only mentioned these 'conditions' in passing, as extra equations to be satisfied after the Lagrange Equations have been solved. But there is a way of incorporating the condition equations from the start - using the method of Lagrange multipliers. We explain the method in mathematical detail in Appendix A6.4.

[45] In A 6.3 we have used a slightly different derivation to Lanczos in order to emphasize the link between E-conservation and the homogeneity of time.

The mathematical arguments are pretty, and the method of Lagrange multipliers enables us to solve otherwise insoluble problems, but we are also very impressed by the fact that the method means something physically. Now in Appendix A6.4 we show how the potential energy, V, has the extra conditions, 'λf', added on, and so we end up with a new potential energy, $V^{new} = V + \lambda f$. However Hamilton's Principle applies in the usual way, and is sensitive to the whole of V^{new} in one go, and pays no heed to its separate components, 'V' and 'λf'. The important physical meaning to extract is that *no ultimate distinction can be made between potential energy and external conditions.*[46] This ties in with what we have claimed earlier (Chapters 4 and 5) - that 'constraints' and 'kinematic conditions' are due to forces, and these forces are ultimately indistinguishable from the applied forces (arising from the potential energy function). It also ties in with the insight derived from d'Alembert's Principle (Section 5.4): there is no ultimate difference between an applied force and a 'fictitious' force (say, due to an accelerating reference frame). Finally, as we just learned in Section 6.6, Hamilton's Principle is sensitive to the whole of $T - V$ and not to T and V separately. We can therefore expand our previous remark and state that: *no ultimate distinction can be made between potential energy, kinetic energy, and external conditions.* We have come a long way from Newton's absolute external force, causing an absolute acceleration, relative to an absolutely stationary and passive Space, and an absolute and passive Time.

There is even one more thing to say about the physical implications of 'condition equations'. In Chapters 3 and 4 we learned that while the degrees of freedom are slippery to define yet they relate to the very essence of the given problem. However we also know that when there are conditions between coordinates, these coordinates are no longer independent of each other, and so *the number of degrees of freedom has been reduced.*

Optional extras

For n generalized coordinates and m condition-equations the number of degrees of freedom is reduced from n to $n - m$. This can be demonstrated geometrically: the n generalized coordinates move within an n-dimensional abstract 'space' (the configuration space), however the m conditions reduce this to a more restricted $(n - m)$-dimensional 'space' within the original 'space'.

[46] (over sufficiently small length- and time-scales)

Although we have indicated that the method of Lagrange multipliers requires *functional* relations between coordinates, actually, it can also be used in the case of 'non-holonomic' conditions, such as: the walls of a gas container; a ball rolling on a surface; a spinning top; and others.[47]

Finally, in a totally original analysis, Lanczos[48] shows that the Lagrange multipliers account for *microscopic* effects (the microscopic reactive forces that maintain the given constraints). The consequential motions, being microscopic, are 'hidden', and they fluctuate in time (are time-dependent), even while the macroscopic system may be conservative. Hertz (end of Section 6.6) wondered whether such 'hidden' motions were the underlying source of all force. In other words, he speculated that kinetic energy (rather than force, or potential energy) was *the* primitive element in physics.

6.9 Symmetries and conservation laws

There is an important kind of 'condition' which brings in great physical insight: the condition known as symmetry. If a system displays symmetry with respect to a given coordinate, q_i, then this means it remains unaltered after a 'small' change, Δq_i, in that coordinate.[49] Equivalently, if the system is symmetrical with respect to a certain q_i then it doesn't depend on that q_i, that is, $\partial L/\partial q_i = 0$ and q_i is known as an 'ignored' or 'absent' coordinate. (Note that L still depends on $\dot{q}_i$ - or else we would have to concede that the system has one less degree of freedom than originally anticipated.)

For example, if a merry-go-round is symmetrical about its axis of rotation then the system does not identify any special angles, or, in other words, $q_i = \theta$ doesn't occur in L. We say that space is 'isotropic' (the same in all directions). Likewise, if a brick can be displaced through a small translation, Δx, and nothing in the world changes as a consequence, then the system does not identify any special positions, or, in other words, $q_i = x$ doesn't occur in L. We say that space is homogeneous (the same at all points).

[47] Goldstein, H, *Classical Mechanics*, 2nd ed, page 12; Lanczos, page 146.

[48] Lanczos, pp 143–5.

[49] This is a kind of symmetry known as a continuous symmetry, as opposed to a non-continuous symmetry such as the symmetry following reflection in a mirror.

Because of the Lagrange Equations, something special happens in the case of these symmetrical systems. If, say, it is the kth coordinate that is ignorable then L does not depend on q_k and so $\partial L/\partial q_k = 0$ but the Lagrange Equation for q_k then implies:

'absent' q_k,

$$\frac{d}{dt}\left(\frac{\partial L}{\partial \dot{q}_k}\right) = 0 \quad \text{and thus} \quad \frac{\partial L}{\partial \dot{q}_k} = \text{constant} \tag{6.21}$$

We shall, in the next chapter, learn to identify $\partial L/\partial \dot{q}_k$, with the 'momentum', p_k, associated with q_k, but for now we note the important result: *a symmetry leads to a conservation law*; a system with translational symmetry leads to a conservation rule for momentum; and a system with rotational symmetry leads to a conservation rule for angular momentum (Noether's Theorem, section 6.7).[50]

Everything we have just said for symmetries with respect to some position coordinate or other, q_i, applies equally well for a symmetry with respect to the time coordinate, t. In fact, we have met this case already. In Section (6.7) we considered a Lagrangian with no explicit dependence on time, and found that a symmetry - invariance of the system after a translation in time - meant that a conservation law emerged: the law of the conservation of energy. We can also argue this another way: as t doesn't occur explicitly it is an 'ignorable' variable and may be eliminated from L; and as a consequence of carrying out this elimination a 'condition equation' is introduced - the very condition that energy is conserved (the maths is shown in Appendix A7.5).

Carl Jacobi (1804–51) tackled the problem of particle motion in this reduced form (that is, with the time coordinate eliminated[51]) and this led to Jacobi's Principle in which $\int mvds$ must be minimized. (In other words, Jacobi had returned to Maupertuis's Principle of Least Action from a hundred years earlier - see Chapter 2.) What Jacobi thereby determined was the path or 'geodesic' of a particle through space,

[50] (Note that, as Lagrangian Mechanics is always a local theory, so these rotations and translations must be 'small'.)

[51] It's a bit more complicated than straightforward elimination: first the time is re-branded as the $(n + 1)$th position coordinate, $t = q_{n+1}$. The time can therefore no longer serve as the independent variable, and all $(n + 1)$ 'position' coordinates are given instead as functions of some new parameter, say, τ. In so doing, it is found that the new Lagrangian depends on $\dot{q}_{n+1}$ but not on q_{n+1} (the differentiation is with respect to τ). Thus q_{n+1} is an ignorable coordinate, and may be eliminated.

but - as befits a model in which time has been eliminated - he determined nothing about the *rate* at which the particle moves along this path. (This motion in time can be reconstructed afterwards by imposing the condition that energy is conserved.) Now in the special case where energy is conserved and there are no external forces (no V), it turns out that the path is straight, and the particle must move along this path with constant speed. Does this sound familiar? Yes - we have recovered the famous 'law of inertia', asserted (in so many words) by Leonardo da Vinci, Galileo, Descartes, and Newton: "A free particle moves [solely under its own inertia] in a straight line with constant speed".

Even more impressive, Jacobi's Principle can be extended and then leads to a *generalized 'law of inertia'*: we take the free 'particle' to be the C-point of a mechanical system of arbitrary complexity and in n dimensions. This C-point moves along a world-line in an n-dimensional Riemannian space - and we find that the paths are the straight*est* lines in that Riemannian space - the 'geodesics' (see Section 3.6 for a mention of Riemannian spaces). We give two examples: (1) a particle is constrained to stay on a given two-dimensional curved surface, with no external forces; (2) in Einstein's Theory of Gravitation, a planet moves in a Riemann space of four dimensions - spacetime - and its motion is governed by the 'law of inertia' with *no* external force of gravity. The only difference between these two examples is that in (2) the curvature of space is an actual property of the physical world,[52] and not a consequence of this or that constraint.

Let us return to our discussions at the beginning of this section - to a consideration of symmetry. We can argue that the 'law of inertia' is upheld because of certain fundamental symmetries: the particle moves in a straight line because space is homogeneous and isotropic - there is no reason for it to prefer left or right; and it moves at constant speed because, in addition, time is homogeneous and isotropic[53] - there is no reason for the particle to move faster or slower. But Lagrangian Mechanics is better: yes, these symmetries may apply, but there is no claim that they extend infinitely far. Even more important, the particle follows the 'straightest' path not only because of local symmetries, but

[52] (the physical world that actually has a large gravitating body, the Sun, about which the planet orbits.)

[53] We must be careful: macroscopically, time does not seem to be the same forwards and backwards; and the Big Bang seems to flag one time as more special than the rest.

because the 'straightest' path is also the 'shortest' path (between given end-points). And finally, the criterion 'shortest' path is more general - it still holds even when there are no symmetries, that is, even when there are external influences, and even when these external influences change in time.

6.10 Conclusions

The basic elements of Lagrangian Mechanics are the generalized coordinates, $\{q_i\}$, the Lagrangian function, $L = T - V$, and the time, t. For each new problem, or even for each new modelling of any one problem, the $\{q_i\}$ are chosen afresh, and the functions T and V may change; and yet we have the remarkable finding that *the Lagrange Equations always have the same form*:

$$\frac{d}{dt}\left(\frac{\partial L}{\partial \dot{q}_i}\right) - \frac{\partial L}{\partial q_i} = 0 \quad i = 1, 2, \ldots, n$$

The invariance of these equations,[54] and their near-universal applicability across the whole of physics, inspires our awe:

> "Was it a God who wrote these signs
> Which soothe the inner tumult's raging,
> Which fill the lonely heart with joy
> And, with mysteriously hidden might,
> Unriddle Nature's forces all around?"

Goethe's Faust[55]
(Faust gazing at the Macrocosmos)

Why are the Lagrange Equations invariant? In view of our discussion in Section 3.6 about differential geometry, that is, the *absolute* calculus, it is not surprising that the true invariant things are connected with infinitesimal changes of one quantity relative to another, that is, with partial differential equations. Furthermore, the reason why these Equations are invariant is that they derive from an 'extremal' condition. But geometry

54 (when taken as a whole, that is, all n equations together)
55 The quote was used in a different context by Lanczos, page 161.

alone cannot tell us which invariant topological features will have *physical* significance; it is only the physics that can answer why it is exactly 'action' that must be minimized.

There is no general, turn-the-handle, way of solving the Lagrange Equations, but by wisely choosing the coordinates - taking full advantage of any prior knowledge of super-structures, symmetries, constraints, and kinematic conditions - we can make the solution easier to obtain, or make a problem tractable as opposed to intractable. We can argue this even more strongly - the increased tractability, mathematically-speaking, means that our wiser modelling of the system actually has more physical meaning - in other words, there really are such things as flexing beams, capacitors, spinning tops, wires with currents in them, gyroscopes, the Solar System, smoothly flowing rivers, and so on, and these things cannot, in general, be built up from the even simpler elements of particles, and forces-between-particles. Wise modelling (incorporating physical knowledge into the system) also confers many computational advantages to Lagrangian Mechanics over Newton's force-mechanics. We have no need to determine the constraint- or reaction-forces (the tension in the cord, the reaction at a pivot, the internal forces maintaining the rigidity of a beam, and so on), or to determine the acceleration (a tricky vector quantity) for each particle in the system. We have only to deal with the *single scalar function*, L, and this function then *determines the entire dynamics of the given system.*

We still have the overarching requirement that Einstein's Principle of Relativity is to be upheld. This is to ensure that the same events will be present, regardless of the reference frame. Consider our earlier example of an electron being attracted to a wire: yes, this 'event' is the same in both the moving and stationary reference frames - the electron is always attracted to the wire. But what of the speed of approach of the electron, and the electron's mass, momentum, and kinetic energy - must these all be unchanged between reference frames?[56] It turns out that the *only* dynamic[57] system-property that is guaranteed invariance is the 'least δ(total action)'.

Now L is made up of the energy functions, T and V. We have made a good case for T having the fundamental form of a 'massy inertial factor'

56 The answer is that none of the above are guaranteed invariance (the mass is only invariant if it is the 'rest mass').

57 There can also be non-dynamic properties of the system that are invariant, like, for example, the total electric charge.

multiplying 'some-function-of-$\dot{q}^2$', but T may also include a constant term, and terms which are linear in the speed; and V usually depends just on position, but may also include a constant term, and terms which are linear in the speed. There is evidently much overlap in the definitions of T and V, so what then is the fundamental difference between them? It is well known that T relates to motions, while V relates to configurations, but, even more fundamentally, T relates to the energy of *individual components*, while V relates to *whole-system* energies. However, there is still the possibility of ambiguity between the two. For example, in earlier discussions (Sections 6.5 and 6.6), we pondered about how to classify rest-mass, being, on the one hand, a static store of energy (so it could be part of V), and, on the other hand, part of the identity of an individual component (so it could be classified as T). This ambiguity is consonant with the fact that L is sensitive to the *whole* of $T - V$, rather than to T and V taken separately. This allows for an interplay between T and V, an interplay that is borne out in nature, and that looks forward to modern physics (field theory, statistical mechanics, quantum theory, gravitation theory, and other disciplines). This is in contrast to Newtonian Mechanics, in which a particle is a particle is a particle - it never affects the space (Space and Time) that it inhabits.

The interplay between T and V occurs through time and not just at one time. So, time has 'configurational' aspects, and may sometimes be treated as a quasi position coordinate (for example, time may be treated as the $(n + 1)$th 'position' coordinate - see the footnote in Section 6.9). Thus we see, in Lagrangian Mechanics, a softening of the formerly sharp distinction between time and space. 'Time' always remains special[58] but, nevertheless, it is sometimes considered on an equal footing with the other 'position' coordinates - and in this way Lagrangian Mechanics foreshadows Einstein's Relativistic Mechanics, in which space and time are no longer independent of each other, but form a continuum known as spacetime.

The previous two paragraphs show us that Lagrangian Mechanics, and not Newtonian Mechanics, survives the transition to Einstein's Theories of Special and General Relativity. This is because Variational Mechanics (which includes Lagrangian Mechanics) is more 'philosophically correct'. Newtonian Mechanics postulates the prior existence of an absolute Space and Time whereas Einstein's Gravitation Theory takes

[58] (Feynman R P, *Lectures on Physics,* Vol I - the same (lost) reference as in Section 3.2.)

the existence of the masses as givens; this is more correct when we reflect that all our observations really have been made in the presence of large gravitating masses. Also, the Variational Mechanics only makes *local* claims (and then the local pieces are joined together, with the condition that the joins must always be smooth). This also is more correct when we remember that 'far away' and 'long ago' are always conjecture, we can never have direct experience of them.

We have claimed near-universal rather than universal applicability for Lagrangian Mechanics - when does it fail? First, the method of Lagrangian Mechanics relies on everything being in functional form (see Sections 3.7 and 6.1). This requirement isn't always met, chiefly in the case of dissipative or frictional effects. However, these dissipative effects are due ultimately to microscopic interactions and, as Lanczos suggests, if we could obtain the functional forms from quantum theory, then variational methods would be applicable after all.[59] Second, we must admit that most real-life problems are just too complicated for *any* analytical mechanics,[60] whether Newtonian or Lagrangian (but Hamiltonian Mechanics, Chapter 7, will rescue some of these problems); then we must resort to curve-fitting, simulations, and other numerical methods.

In summary: Lagrangian Mechanics is a local theory; the time and position coordinates are not required to be independent of each other; T has inertial attributes; in fact, T *and* V have inertial attributes (and this accords with Einstein's finding that the mass-energy equivalence applies to *all* types of energy); the sharp division between geometry and external conditions has broken down; and 'energy' rather than 'force' is the true determinant of what happens. What we have not mentioned is that the Variational Mechanics also shows the way into Quantum Mechanics. This will be looked at in Chapter 7, which charts the next great advance, brought about by Hamilton.

[59] Note that introducing 'polygenic' frictional forces, or non-holonomic (microscopic) conditions, introduces a right-hand side to the Lagrange Equations, see Appendix A6.5 and Lanczos, pp 146–7.

[60] 'Analytical' implies that every aspect can be modelled mathematically, and then the equations can be solved.

Figure 7.1 Nicolas Poussin, *A Dance to the Music of Time*, c. 1640, by permission of the Trustees of the Wallace Collection, London.

A metaphor for the perpetual dance of the p s and q s in phase space.

7

Hamiltonian Mechanics

"A dance to the music of time"
Nicolas Poussin, c. 1640

7.1 Introduction: ask less from more

The outstanding achievements of Lagrange are still not the last word in mechanics, and it was an Irish mathematical prodigy, William Rowan Hamilton, who, in the nineteenth century, took mechanics to its highest form. Hamilton was in awe of Lagrange, referring to him as a Shakespeare, and to the *Mécanique analytique* as a scientific poem; it was this work which attracted him to the topic of mechanics. Hamilton understood that even if the equations of motion were sometimes too difficult to solve one could nevertheless obtain important qualitative information - but only if one used the right choice of variables. His crucial advance was to discover what were the true, most telling, variables of mechanics.

We are familiar with the fact that the choice of variables (coordinates) can make all the difference to the tractability of a problem in mechanics. For example, a lever can be modelled as a near infinity of atoms, or as a 'lever arm' with just one position coordinate (the angle, θ) and the condition of 'rigidity'. We are inclined to think that the second version is an improvement over the first, but Hamilton realized that it is not always best to have the sparest, most economical description; sometimes even an increase in the number of coordinates can lead to greater insights.

Specifically, Hamilton brought in a *doubling* of the number of coordinates in any mechanics problem. This was no mere doubling of the number of dimensions (as would be the case in going from, say, a class of 25 children to a class of 50 children) but a doubling in the 'dimensionality' of the problem (as in going from 'children' to 'boys' and 'girls'). This analogy is useful but too simple, it doesn't demonstrate Hamilton's further requirement - that the two kinds of variable must be *dynamically*

The Lazy Universe. Jennifer Coopersmith, Oxford University Press (2017).
 DOI 10.1093/acprof:oso/9780198743040.001.0001

related. A better analogy[1] is the example of a semiconductor crystal: an electron may occasionally be missing from a given lattice site, and then there will be a 'hole'. The positions of the holes are necessarily implied by the positions of the electrons but in practice the holes take on a life of their own, and it is useful to map both the electrons and the holes independently from each other. Another example has to do with the rings of Saturn. The rings are made up of particles, and for each particle we could tabulate both its radius and its speed even though if one is known the other is automatically determined. The data-table can then be trawled through as many times as necessary: we may, for example, look for all particles orbiting beyond a certain radius, and then on another occasion look for all particles with a speed below a certain threshold. However, Hamilton's Mechanics is so much more than a matter of picking out trends in a spreadsheet. Consider the suggestive analogy of a picture, which, as we know, is made up from an array of pixels. The amount of input data is doubled up by the simple expedient of viewing the picture with two eyes as opposed to one. In especially contrived stereographic pictures this doubling-up reveals previously hidden objects and a hidden depth. Likewise, we shall find that Hamilton's doubling of variables leads to hidden depths of understanding.

7.2 The optical theory: extraordinary genius

We have learned in the previous chapters that Lagrangian Mechanics can be cast as a question in geometry (an 'extremal' condition in an abstract space). Similarly, Hamilton's big advance was also inspired by geometry. When a youth of only seventeen years, Hamilton was musing on the problem of mathematically describing an 'optical system' - a system comprising a tight bundle of light-rays, starting at a common source, and then passing through various lenses and mirrors. His aim was to find the most general description possible of the optical system - one mathematical formulation that would serve whatever the arrangement of lenses or mirrors, and a formulation that was not to depend on the physical nature of light, that is, whether light is a wave or a particle. Hamilton came to the realization that as the rays pass through a system of this, that, or the other lenses and mirrors there are certain *geometrical*

[1] Most of the analogies we employ from now on can, in fact, be cast into problems of Hamiltonian Mechanics.

properties that remain unchanged: the ray-tips define a surface - a 'surface of simultaneous arrival time' - and if the rays leave such a 'surface' at right angles, then they will leave all subsequent 'surfaces of simultaneous arrival time' at right angles. (The same geometry applies for arrivals, that is, if the rays arrive at such a 'surface' at right angles, then they will arrive at all subsequent 'simultaneous surfaces' at right angles.) We call this geometric property the 'ray property'. Given the exceedingly fast speed of light, this remarkable observation was available only to Hamilton's mind's eye, not to his actual eye.

Hamilton knew of Descartes's epoch-changing discovery, made almost two hundred year earlier, that geometrical properties (Euclid's axioms about planes, triangles, parallel lines, and so on) could be described purely algebraically, that is, by algebraic[2] *functions of coordinates*. (It was, in fact, Descartes who brought in the very idea of 'coordinates'.) This led Hamilton to wonder whether he could explain the 'ray property' purely by an algebraic *function of coordinates*. In mathematics, the geometric property of being a surface can indeed be expressed as a function: $f_{surface} = 0$, where $f_{surface}$ is some algebraic function of the relevant coordinates. But Hamilton also appreciated that, because of Fermat's Principle of Least Time, then one such 'surface of simultaneous arrival time' is *functionally linked* in some way to any subsequent 'surface of simultaneous arrival time'. Now Fermat's Principle dictates that the 'distance' (in time) between these surfaces has to be a minimum - but can this 'minimum distance' itself be reformulated as an algebraic function? Even for a mathematician of Hamilton's calibre, this was a complicated thing to do. Hamilton was not daunted (he was barely 18 years old) and he looked at this complexity from an astoundingly audacious angle, as we now explain.

Using the Principle of Least Time, the 'least-time distance' for a light ray travelling between specified initial and final positions, q_i and q_f, can be determined. Having done this, the whole process can be repeated but for a slightly different choice of initial and final positions, say, q_i' and q_f'. A slightly different 'least distance' will ensue. This evaluation can be repeated again and again, in each case for slightly different end-state coordinates, and in each case yielding slightly different 'distances'. The audacious question Hamilton asked was: could this 'least

[2] An algebraic function is the solution to a polynomial equation. It is defined by a finite series of operations - for example, adding, subtracting, multiplication, and 'to the power of' are allowed, but trigonometric operations are not allowed.

distance' be a function, perhaps even an *algebraic function*, of these end-state coordinates? The answer he found was - yes, the 'distance' was an algebraic function, let's call it, $f_{distance}$, of the end-state coordinates, and the end-state coordinates alone. (Readers may be reminded of other scenarios in physics where there is function that depends only on end-coordinates: for example, a conservative potential energy field, or a 'function-of-state' in thermodynamics.)

Thus Hamilton came to the realization that there were two overarching algebraic functions that completely described the system of rays: first there was the function that determined the surface of simultaneous arrival time, $f_{surface}$, and then there was the function, $f_{distance}$, that was the 'distance' between two such surfaces. So far, this was genius of an ordinary kind; next came the hallmark of extraordinary genius - for Hamilton had to decide what coordinates to use, and then to find what the functions actually were.

7.3 Hamilton's Mechanics - the right coordinates

The problem was first solved for the case of optics. Hamilton's work in optics, "Theory of Systems of Rays",[3] was published in 1828 and in it he introduced his 'characteristic function', so-called because it completely characterised the given optical system (the arrangement of mirrors and lenses, and the refractive indices of the media). This 'characteristic function' *is* the function, $f_{distance}$, and it depends only on the start and end position-coordinates of the light rays. In addition, Hamilton also determined the function, $f_{surface}$, (for example, for light rays which arrived in parallel at a concave mirror and were then focused to a point, $f_{surface}$ was the function known as the 'caustic surface').

It would be over ten years before Hamilton's work on mechanics appeared yet the seeds were already evident in the earlier Optical Theory, as Hamilton understood that the geometric 'ray property' arises in *any* system governed by a minimum principle. (It was later proved that the reverse is also true: a minimum principle arises from the 'ray property'.) What was the minimum principle operating in mechanics? Hamilton knew it had to be a principle concerning action rather than time - after all, it was he who had brought in Hamilton's

[3] Hamilton W R, "Theory of Systems of Rays", *Transactions of the Royal Irish Academy*, vol 15 (1828) pp 69–174.

Principle (see Chapter 6), a generalization of Lagrange's Least Action Principle, itself an overhaul of Maupertuis's Least Action Principle. But what is the equivalent of a ray when it comes to mechanics? Hamilton's answer: it is the worldline of a single 'whole-system' point moving in an abstract multi-dimensional space. We have met this before: the whole-system point is the C-point (Chapter 6), and its worldline in configuration space is equivalent to a 'ray'. In summary, in optics there are light rays in everyday 3-D space, influenced by lenses and mirrors, and subject to a principle of least time; while in mechanics there are fictitious C-points moving along 'rays' in the multi-dimensional abstract configuration space, and subject to a principle of least action: and in both optics and mechanics the 'rays' exhibit the 'ray property'.

We must now consider what the functions, $f_{surface}$ and $f_{distance}$, are in mechanics, and what coordinates should be used. Despite certain inescapable differences[4] between optics and mechanics, the former is much simpler to visualize, and therefore offers valuable insights into mechanics. In the optical system, the 'simultaneous surface' really is a 2-D surface in everyday 3-D space.[5] In mechanics, the 'surface of common action' is a complicated thing known as a hyper-surface,[6] and yet by analogy with light it still has certain geometrical properties - the 'ray property' - as if it were an ordinary 2-D surface: in optics $f_{surface}$ has *two* dimensions; in mechanics, $f_{surface}$ has *two* dimensionalities (the terminology was explained in Section 7.1). This means that in mechanics $f_{surface}$ depends on *two* coordinates for each particle on the 'surface'. Finally, going back to the introduction section of this chapter, we suspect that these two surface-coordinates for one particle may be dynamically related to each other. So much for $f_{surface}$, for the moment.

In mechanics, the function $f_{distance}$ links two 'surfaces of common action' together and so, like $f_{surface}$, it also must have a dimensionality of two (we have just one mechanics problem, and the dimensionality of a problem can't change halfway through the analysis). However, this dual dimensionality comes about in a new way. We remember that $f_{distance}$ arises from an integral, that is, an integral between *two* end-states.

4 For example, in optics, the light rays arrive at their 'geometric surface' at the same time, whereas, in mechanics, the C-points arrive at their 'surfaces' at different times but at common values for action.

5 (for example, for a point light-source, a spherical surface centred on this source receives the rays simultaneously and at right-angles)

6 A hyper-surface is a surface in more than two dimensions.

The requirement for a dimensionality of two is therefore most easily and suggestively satisfied if we adopt these start- and end-'positions' for each 'ray' as the dual coordinates for $f_{distance}$ in mechanics.

To sum up, we have made a good case for $f_{distance}$ depending on end-state coordinates, *one* on each of *two* different 'surfaces'; and a good case for $f_{surface}$ depending on *two* dynamically-related coordinates, both on *one* 'surface'. It now remains to ascertain *what* exactly these two dynamically-related surface-coordinates will be. (Note that, in this mechanics scenario, there is also the possibility of $f_{distance}$ and $f_{surface}$ depending on the time, t.)

We are satisfied that $f_{distance}$ depends on the two end-positions for each ray, but where will the duality come from in the case of $f_{surface}$? We cannot hope to bootstrap the modelling of physics from plausibility arguments alone - at some point a daring visionary has to come along and show us the way (and then their theory must make predictions, and then these must be tested against experiment). Nevertheless, there is just one more 'plausibility argument' that we can follow. Remember that in our explanation of the variational calculus (Section 3.7, Chapter 3) we stated that the integrand function, L, has *two* kinds of argument,[7] q and $\dot{q}$, (and the reason for this peculiarity was explained in Section 6.4). This seems like a gift, too good to ignore - might we not take the q_i and the $\dot{q}_i$ and employ them together as our *two* coordinates per particle, in the definition of $f_{surface}$? Yes, this is exactly what we'll do - but there's a subtle point that needs explaining. The arguments of a function may be split into separate categories (the notational convention is to separate each category by a semi-colon) but when reassigning categories - say, shifting the $\dot{q}_i$ s (the speed coordinates) into the same category as the q_i s (the position coordinates) - then we must be aware that the $\dot{q}_i$ s will change their nature and turn into position coordinates. A good allegory comes from a cutlery drawer: the forks, knives, and spoons, are all stored in different compartments, but when the forks are shifted, say, to the spoon-compartment, then the forks must be used *as if* they were now spoons.

The daring visionary was, of couse, Hamilton, and he realized he could double the number of coordinates at a stroke by transferring the

[7] Not counting the time, t. Note that we're now using the term 'argument' to mean 'argument of a function'.

$\dot{q}_i$ s into the same category as the q_i s. However in order for this stratagem to work he knew he had to disguise the $\dot{q}_i$ s making them not look like speed coordinates anymore. He accomplished this by performing a transformation as follows:

Lagrange's description			Hamilton's description	
q_i,	position	$\mapsto$	q_i,	position
$\dot{q}_i$,	speed	$\mapsto$	p_i,	position
t,	time	$\mapsto$	t,	time

with p_i given by:

Defining equation for conjugate 'momentum', p_i,

$$p_i = \partial L/\partial \dot{q}_i \qquad \text{for all } i = 1 \text{ to } n \tag{7.1}$$

In (7.1), L is 'the Lagrangian', and it is still a function given in terms of the original untransformed coordinates, q_i and $\dot{q}_i$. There may also be an explict dependence on time and, if so, the t is the same in Lagrange's and Hamilton's descriptions. This is still a bit baffling - how are the $\dot{q}_i$ actually transformed into the p_i? Well, as the Lagrangian, L, is a function of q_i, $\dot{q}_i$, and possibly t, then $\partial L/\partial \dot{q}_i$ is likewise some function of q_i, $\dot{q}_i$, and possibly t, and so p_i is also some function of q_i, $\dot{q}_i$, and possibly t. But this p_i-function can then be rearranged ('inverted'[8]) to yield $\dot{q}_i$ as a function of q_i, p_i, and possibly t. This $\dot{q}_i$-function, is, finally, the function we require to carry out the transformation: wherever we find a $\dot{q}_i$ in the original Lagrangian we replace it by the $\dot{q}_i$-function of q_i, p_i, and possibly t, and so the transformed Lagrangian itself becomes a function of q_i, p_i, and possibly t. (See Appendix A7.1 for some worked examples.)

In the new scheme, we now have two 'position' coordinates for every one old position coordinate: $(q_i) \mapsto (q_i,\ p_i)$. We need to explain the words used in (7.1). The adjective 'conjugate' is employed to remind us that each q_i is linked to one corresponding p_i (and vice versa). This 'conjugate pair' are for one and the same particle, i, and are dynamically related (they are connected via equation (7.1)); there is no special relationship linking, say, q_2 and p_8.[9] The descriptor, 'momentum', is

[8] We assume this is always possible - the Jacobian must be non-singular.

[9] However, through the presence of L in (7.1), all i-values may enter into all the transformation equations.

employed for the following reason: we remember that the speed coordinate, $\dot{q}_i$, usually occurs in L in the form $\frac{1}{2}m_i\dot{q}_i^2$ (Section 6.5.1), and then $\partial L/\partial \dot{q}_i$ turns out to be $m_i\dot{q}_i$ - but this is none other than the standard definition for momentum. However there is more that can be said. Momentum is more than just speed re-scaled, it has a massy factor, and Hamilton's transformation equation, (7.1), allows for this massy factor to enter in many ways - for example, the mass need not be constant, it could depend on speed, spatial distribution, or time. Thus, even when L has a non-standard form, equation (7.1) still applies, and then p_i is the *generalized* momentum (see Chapter 3 on generalized coordinates). Note further that although the p_i s are called 'momenta', they are to be considered as position coordinates (remember the cutlery drawer), of the same status as the q_i coordinates, and so altogether, for each 'particle', i, we end up with *two* generalized *position* coordinates, (q_i, p_i).

7.4 The canonical equations

We have arrived at the required duality, but, as we have seen, Hamilton has not plucked his new (q_i, p_i) coordinates out of the air, he has obtained them (via (7.1)) from Lagrangian Mechanics, and this is because he wants to continue to use the Principle of Least Action as the founding principle of his new mechanics. Therefore, we have the least action condition (condition (6.11)) still at the heart of the physics problem. In Hamilton's Mechanics, we use this condition as before, but only after replacing the original Lagrangian by our newly transformed Lagrangian, L^{new}. So far, so unexceptionable - but then something quite astonishing happens (even while it will sound like a succession of "Just So"[10] stories). The new Lagrangian, L^{new}, is split into T^{new} and V^{new}, as usual, but then it just so happens (following on from our transformation equations, (7.1)) that L^{new} always ends up having a rather special form:

$$L^{new} = T^{new} - V^{new} = \left(\sum_i p_i\dot{q}_i\right) - V^{new}(q_i, p_i; ; t) \tag{7.2}$$

That is to say, it turns out that T^{new} always has the form $(\sum_i p_i\dot{q}_i)$, and it turns out that V^{new} has no $\dot{q}_i$ s or $\dot{p}_i$ s in it (and this seems to confirm

[10] R Kipling, Just So Stories for Little Children, Macmillan & Co. (1902).

our calling it a *potential* energy function[11]), and, furthermore, it turns out that V^{new} is a purely *algebraic* function of the q_i s and p_i s, and possibly t. Is V^{new} the function, $f_{surface}$, that Hamilton was seeking? Not quite, but almost.[12] To honour it, V^{new} is given a special name: it is called 'the Hamiltonian' and given the symbol 'H'.

Commentary 1)
To repeat, H is a potential energy function, a scalar algebraic function just of p_i s, q_i s, and possibly t; and T^{new} always has the form $(\sum p_i\dot{q}_i)$. However, we may object, and wonder why the $\dot{q}_i$s are still hanging around (in T^{new}) when, according to the procedure given in Section 7.3, they should all have been replaced by their $\dot{q}_i$-functions - (functions just of p_i s, q_i s, and posssibly t)? There are two answers to this. The first is to argue that we can decide on purpose to normalize T^{new} to this new specific form, and therefore we can choose not to replace *all* the $\dot{q}_i$ s in L^{new} but leave just the necessary ones in place. The second answer is better - less *ad hoc*; it is that L^{new} is in principle a function of $2n$ position coordinates and $2n$ consequential speed coordinates but it just so happens that none of the $\dot{p}_i$ s appear, and it just so happens that the only $\dot{q}_i$s that appear are the ones in $(\sum p_i\dot{q}_i)$:

$$L(q_1, \ldots q_n; \dot{q}_1, \ldots \dot{q}_n;t) \mapsto L^{new}(q_1, \ldots, q_n, p_1, \ldots, p_n; \dot{q}_1, \ldots, \dot{q}_n, \dot{p}_1, \ldots, \dot{p}_n;t)$$

where it just so happens that:

$$L^{new} \quad \text{always has the form} \quad \left(\sum_i p_i\dot{q}_i\right) - H(q_1, \ldots, q_n, p_1, \ldots, p_n; ; t) \tag{7.3}$$

This new Lagrangian can be put into the minimized action condition,

$$\delta\int_{t_a}^{t_b} L^{new}\, dt = \delta\int_{t_a}^{t_b} \left(\sum_i p_i\dot{q}_i\right) - H(q_1, \ldots, q_n, p_1, \ldots, p_n; ; t)\, dt = 0 \tag{7.4}$$

and this will lead to the Lagrange Equations, as usual (except that there will be $2n$ instead of n of them):

[11] In Section 6.5.2, we said that, in the main, the potential energy depends on position- rather than speed-coordinates.

[12] We'll find that H is not identical with $f_{surface}$ as the former is a surface of *energy*, whereas the latter is a surface of common *action*. However, in cases where H is independent of t, then H and $f_{surface}$ are linearly related - see Section 7.7.

$$\frac{d}{dt}\left(\frac{\partial L^{new}}{\partial \dot{q}_i}\right) - \frac{\partial L^{new}}{\partial q_i} = 0 \qquad i = 1, 2, \ldots, n$$

$$\frac{d}{dt}\left(\frac{\partial L^{new}}{\partial \dot{p}_i}\right) - \frac{\partial L^{new}}{\partial p_i} = 0 \qquad i = 1, 2, \ldots, n \qquad (7.5)$$

Now, because of the special form of L^{new} as in (7.3), then $\partial L^{new}/\partial \dot{q}_i$ is always just equal to p_i (and, furthermore, this is consistent with the defining equation, (7.1)), and $\partial L^{new}/\partial q_i$ is the same thing as $-\partial H/\partial q_i$ (as q_i only occurs in the H part of L^{new}), and $\partial L^{new}/\partial \dot{p}_i$ is always zero (as there is no $\dot{p}_i$ in L^{new}), and, finally, $\partial L^{new}/\partial p_i$ is always given by $(\dot{q}_i - \partial H/\partial p_i)$. Making all of these substitutions into (7.5) we arrive at:

$$\frac{dp_i}{dt} - \left(-\frac{\partial H}{\partial q_i}\right) = 0 \qquad i = 1, 2, \ldots, n$$

$$0 - \left(\dot{q}_i - \frac{\partial H}{\partial p_i}\right) = 0 \qquad i = 1, 2, \ldots, n \qquad (7.6)$$

Rearranging, and noting that dp_i/dt may be written as $\dot{p}_i$, we finally arrive at:

$$\dot{p}_i = -\frac{\partial H}{\partial q_i} \quad \text{and} \quad \dot{q}_i = \frac{\partial H}{\partial p_i} \qquad i = 1, 2, \ldots, n \qquad (7.7)$$

These equations are the new equations of motion. They are sometimes called 'Hamilton's Canonical Equations', and we shall find that they are a turning point in mechanics, ushering in a new era. It is not essential to follow the mathematical derivation above; just know that we started from the original Lagrangian Equations (6.13), applied the transformations (7.1), and ended up with Hamilton's Canonical Equations. (The adjective 'canonical', coming from 'canon law' in church proceedings, was coined by the mathematician Carl Jacobi (1804–51) (see Chapter 2), and conjoined forever afterwards with Hamilton's Mechanics as a gesture of respect and admiration. Applied to the equations, 'canonical' means *the* definitive and correct form for mechanics; applied to the p_i s and q_i s, 'canonical' means *the* definitive and correct choice of coordinates.)

In order to understand the advance that Hamilton's Mechanics represents, let us assemble the old and new equations of motion together:

From: **The Lagrange Equations**

$$\frac{d}{dt}\left(\frac{\partial L}{\partial \dot{q}_i}\right) - \frac{\partial L}{\partial q_i} = 0 \qquad i = 1, 2, \ldots, n \qquad (7.8)$$

To: **Hamilton's Canonical Equations**

$$\dot{p}_i = -\frac{\partial H}{\partial q_i} \qquad i = 1, 2, \ldots, n$$
$$\dot{q}_i = \frac{\partial H}{\partial p_i} \qquad i = 1, 2, \ldots, n \qquad (7.9)$$

Once again, the words of Goethe are apt:

> "Was it a God who wrote these signs
> Which soothe the inner tumult's raging,
> Which fill the lonely heart with joy
> And, with mysteriously hidden might,
> Unriddle Nature's forces all around?"

Goethe's Faust[13]
(Faust gazing at the Macrocosmos)
Part I, Scene 1

We have already explained, in Chapter 6, why the Lagrange Equations inspire our awe; let us now explain why Hamilton's Equations command a ten-fold increase in awe.

Hamilton's Equations show how the q_i s and p_i s undergo a 'dance to the music of time',[14] a dance in which, as some q_i s or p_i s increase in value, others decrease in value, but always such as to keep the energy constant (in conservative systems), and always such as to keep the total action minimized, both instant by instant, and over the whole path between 'surfaces-of-common-action'. This 'dance' is governed by *one* function, H, - that is to say, while H is different for different systems (orbiting planets, a statistical ensemble, an electrical circuit, positrons orbiting an atomic antinucleus, a spinning top, juggling pins, a flowing

[13] As quoted by Lanczos page 161.

[14] Nicolas Poussin, A Dance to the Music of Time, c. 1640. (See Chapter 7 quotation and Figure 7.1.)

river, and so on) yet within any one system there is just one overarching function (there is no need for individual functions, H_1, H_2, $\ldots H_n$). Also, H is an algebraic function, and so - in sharp contrast to Lagrangian Mechanics - in order to solve the equations just simple differentiations and substitutions suffices. Also, in Hamiltons Equations, all the time-dependence (the dotted variables) are neatly together in one place (on the left-hand side of the equations) while all the algebraic operations are neatly together in one place (on the right-hand side of the equations). Also, this time-dependence is only to first-order (there are single dots, $\dot{q}$ and $\dot{p}$, but no double dots, $\ddot{q}$ and $\ddot{p}$, or higher orders), and also, the equations are linear in their time-dependence (we have $\dot{q}_i$ and $\dot{p}_i$ but no $\dot{q}_i^2$, $\dot{p}_i^2$, $\dot{q}_i^3$, $\dot{p}_i^3$, and so on).

But it is not principally for the sake of increased mathematical ease that we are awe-struck, but rather for the sake of increased physical insight (although the two are evidently related). In particular, the simple mathematical structure of Hamilton's Equations makes them especially able to bring out physically telling symmetries and conservation theorems, such as the conservation of momentum, the conservation of angular momentum, and the conservation of energy (see Section 7.6). More than this, Hamilton's Mechanics brings in not only a generalized definition of momentum (equation (7.1)), but a more generalized and fundamental definition of energy: the kinetic energy is normalized to a special form, and it is no longer necessary that it is 'quadratic', or that the potential energy is velocity-independent, or speed-independent, or time-independent, and it is not even necessary that energy be conserved. Perhaps the single most astounding new physical insight arising out of Hamilton's Mechanics will be the 'wave-nature of particles' and the 'particle-nature of waves' (see Sections 7.7, 7.8, and Chapter 8). We begin to appreciate why Lanczos, and also Schrödinger (Erwin Schrödinger (1887–1961), the discoverer of the wave equation in quantum mechanics, Section 7.8), claim that the Hamiltonian, H, is the most important function in mechanics:

> "The central conception of all modern theory in physics is the "Hamiltonian"..."[15]

[15] E Schrödinger, The Hamilton postage stamp: an announcement by the Irish minister of Posts and Telegraphs, referenced in T L Hankins, *Sir William Rowan Hamilton*, Johns Hopkins University Press (1980), p64 note 7.

Commentary 2)
The minimization of the action integral, (7.4), requires that all the position coordinates are varied independently of each other - but isn't this requirement contravened by the fact that the p_i are related to the q_i (through (7.1))? Answer: we treat the δp-variations *as if* they were independent. This can be done because it just so happens that the minimized-action-condition, $\delta(\text{Action})=0$, is satisfied anyway, irrespective of the δp s. (And the reason for this is that the transformation equation, (7.1), implies that the coefficient multiplying each δp is identically zero, just like the coefficients in the case of Lagrange Multipliers[16]). How this comes about is explained quite simply in Appendix A7.2.

We are still left wondering: surely the differential nature of $\dot{q}_i$ cannot be hidden by re-naming it p_i? Also, surely we can't expect to learn twice as much just by having twice as many coordinates? A partial answer: yes, we have n more coordinates, but we have introduced n more constraint-equations (the very transformation equations (7.1)). A better answer: what saves Hamilton's camouflage 'trick' from emptiness is the fact that, at the same time as doubling the number of coordinates, we have moved the mechanical problem into a totally new space: we have moved from configuration space[17] in n dimensions to a space called 'phase space' (next section) in $2n$ dimensions. These spaces are different from each other and, crucially, they are both abstract - and of utterly different abstract spaces one can ask utterly different abstract questions. This will be explained further in Section 7.5. Before proceeding with this explanation, we give some worked examples of simple mechanics problems, solved using the methods of Hamiltonian Mechanics - see Appendix A7.3.

7.5 A fluid flowing in phase space

It would seem, from Appendix A7.3, that there has been no advantage in using Hamilton's Canonical Equations over Lagrange's Equations over Newton's Equations - so why have we bothered? A short allegory will help to explain the different aims of Lagrangian Mechanics and Hamilton's Mechanics - and explain why we do bother.

16 See Appendix A6.4.

17 (See Section 3.5, Chapter 3.)

Imagine that we are keen on golf and want to improve our stroke. On Saturday, we are at the tee of hole number 18, we have selected our golf club, have an ample supply of identical golfballs, and proceed to hit 100 balls towards flag number 18. Exhausted, we walk over to the putting green and count up the number of balls we find there. The next day (Sunday) we again drive 100 balls, but just as we're about to walk to hole 18 it starts to rain and we head, instead, for the clubhouse, where tea and scones awaits us. Fortunately, our companion used his smartphone to take photographs of each drive, and the phone has been programmed (using Lagrange's Mechanics) to calculate the trajectory of a golfball, knowing the angle and speed at which it leaves the golf club, and so determine whether the given ball makes it to the putting green.

One might think that there's not much to choose between the methods employed on Saturday and then on Sunday (apart from the fact that in one case we had need of a clever computing device) but there's a world of difference: on Saturday, we count the number of balls on the green *after* their arrival; on Sunday, we calculate the whole trajectory of a given ball and so we know whether the ball arrives, and *when*. We can say that Saturday's and Sunday's results occur in different 'spaces'. In the 'Sunday space', we can reconstruct the entire history of each and every golfball; in the 'Saturday space', we are happy to forego this detailed knowledge because we really just want to know what proportion of our drives do in fact make it to the putting green. We could also investigate other questions of a general nature, such as whether any golfballs at all will make it through a certain gap in the trees, and what overall difference the choice of golfclub makes, and so on. (If we need to know more about one specific ball or another, this more detailed knowledge can be reconstructed afterwards, if we supply the appropriate extra data.)

This allegory nicely demonstrates the sorts of differences we find between phase space (the 'Saturday space') and configuration space (the 'Sunday space'). In phase space we obtain qualitative information, about more golfballs, all in one go - we obtain 'less from more'. Before we give some examples, let's first explain how a plot in phase space is constructed.

Phase space

In Hamilton's Mechanics we have a problem in $2n$ dimensions - the n q_is and the n p_is. We wish to picture this graphically, and so, as a matter of convenience, we plot the q_is and the p_is as rectangular coordinates of a $2n$-dimensional space - that is, we plot the q_is and the p_is against straight, perpendicular axes.[18] (In configuration space, Section 3.5, we likewise plotted the q_is against straight, perpendicular axes.) In phase space we can then track the progress of one 'whole-system' C-point along its worldline in this abstract space of $2n$ dimensions. While the C-point is fictitious, we can sometimes choose units such that it mimics the behaviour of a real particle.[19]

An interesting question is: can we 'beat the system'? That is to say, suppose we consider a system with just one degree of freedom, q, and suppose this q really is a rectangular coordinate (for example, it is the horizontal distance travelled by a bullet rather than, say, the θ-coordinate of a pendulum). Then, in this special case, could it be that the world-line of the C-point in configuration space is identical to the actual bullet's trajectory (what you can see with your eyes)? Yes, this could happen - but we can never beat the system as far as phase space is concerned. We might fire an identical bullet again and again, say, 20 times, and plot the 20 individual world-lines on one graph. In the configuration space graph, these 20 individual world-lines could cross each other, like straw in a barn (as it may happen that two bullets with different starting speeds end up going through the same position later on). However, in the phase space graph the 20 individual world-lines could *not* cross each other; on the contrary, they would have to follow similar paths nearby, like streamlines in a fluid. This is because each point in phase space corresponds to a unique[20] pair of coordinates, (p_i, q_i). If it did so happen that two identical bullets were fired with the same velocity, and from the same position, then they would follow the identical 'streamline' in phase space.

It is not possible to draw a $2n$-dimensional space, and so we give some simple examples of two-dimensional phase spaces (that is, $n = 1$). In Figure 7.2, we have a longitudinal spring, fixed at one end, and free to oscillate at the other end.[21]

[18] (Even in 4 or more dimensions, it is possible to give meanings to the terms 'straight' and 'perpendicular'.)

[19] For example, for a system of N free particles, we can define the metric, ds, (see Section 3.6) according to: $T = \frac{1}{2}m(\frac{ds}{dt})^2$ and then set $m = 1$. We then find that the C-point behaves like *one* particle of mass 1 and kinetic energy T, moving in a $3N$-dimensional configuration space.

[20] (but see Figure 7.4 - the explanation is in the nearby text)

[21] Calculated using Scilab, to reproduce Fig. 6 in Lanczos, page 179.

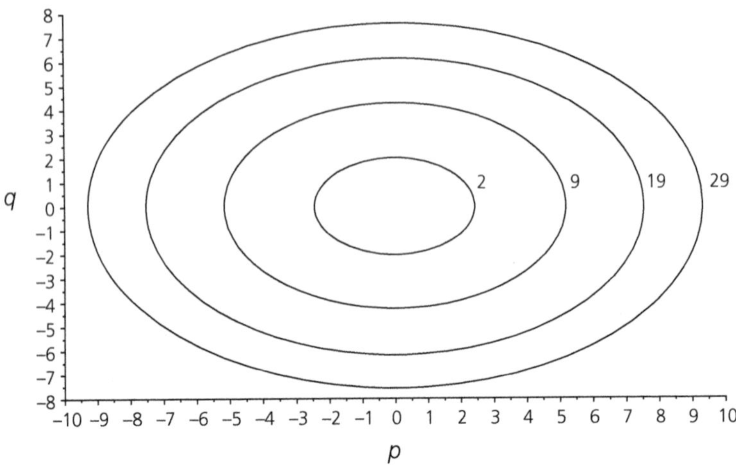

Figure 7.2 An oscillating spring.

The coordinate q measures the extension of the spring, and the coordinate p is a measure of the speed of the free end of the spring. Each elliptical streamline is for the same spring released from a different starting extension. (The number on the streamline is the constant value of action on that streamline.) One can see that whenever the spring is maximally extended, that's when the speed is slowest (p goes through zero).

In Figure 7.3, there is a simple pendulum.[22] The bob has two modes - circulating or oscillating. When circulating, it starts by hanging straight down ($\theta = 0$) and then circulates to being straight up ($\theta = \pm\pi$) and then continues on in the same direction (clockwise or anticlockwise), and eventually reaches hanging straight down again, but immediately continues circulating, and so on, and so on (as we are ignoring dissipation). The circulation streamlines are the open almost horizontal curves at the top and bottom of the figure. When oscillating, the bob has small angular displacements but never reaches $\theta = \pm\pi$. The oscillation streamlines are the closed oval-shaped curves. There is one special streamline, the eye-shaped 'separatrix', that separates circulation from oscillation. For all the streamlines, the speed is at a maximum as the

[22] Figure 7.3 was generated with Scilab using initial parameter-values as in G J Sussman and J Wisdom, *Structure and Interpretation of Classical Mechanics*, MIT Press (2001), Figure 3.4 page 208.

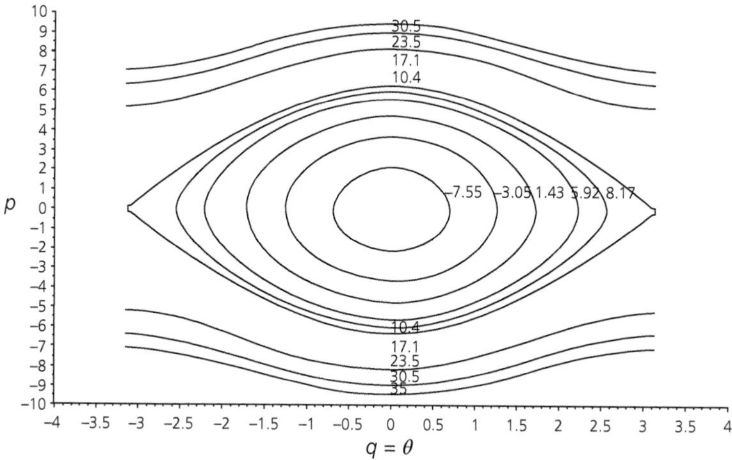

Figure 7.3 Simple pendulum (length 1 m, bob of mass 1 kg, and the acceleration of gravity is 9.8 m s^{-2}.)

bob passes through zero. (As in the previous example, the number on the streamline, or 'iso-action' curve, is the constant value of action on that streamline.)

The final example is for an axially-symmetric spinning top (Figure 7.4).[23] Because of symmetries in the system, we only have to consider the angle of tilt, θ, of the top from the vertical. (The other Euler angles, ψ and ϕ are 'ignorable' - see Sections 6.9 and 7.6. Also, setting $p_\psi = p_\phi$ means that $\theta = 0$ when the top is vertical.) The Figure shows streamlines of p_θ against θ for a given constant rotation, ω, of the top about its symmetry axis. Note that there is a point where the streamlines cross - but, wait a minute, didn't we just say that this could never happen in phase space? The resolution is that, starting from anywhere on this 'figure-of-8' streamline, it takes an infinitely long time to reach the crossing-point…

All in all, we are awed by the fact that the flowing-fluid metaphor is holding strong - even when applied to systems as 'non fluid-like' as an oscillating spring, a pendulum, and a spinning top.

One signal advantage of using Hamilton's Mechanics, and phase space, is that, even where some problems may be too difficult to solve

[23] Figure 7.4 was generated with Scilab using initial parameter-values as in Sussman and Wisdom, Figure 3.7 page 216.

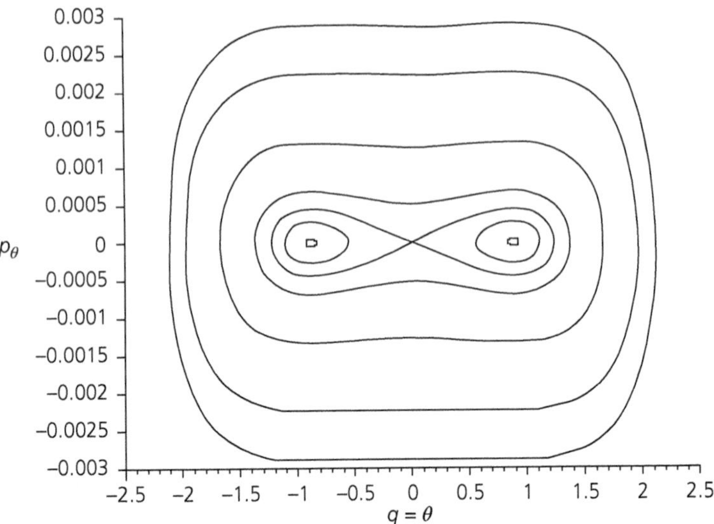

Figure 7.4 Spinning top (calculated using parameters from Sussman and Wisdom, Structure and Interpretation of Classical Mechanics, MIT Press, 2001, Figure 3.7, page 216. θ is the angle between the top's axis and the vertical. The rotation rate is set at $\omega = 90$ rad s^{-1}.)

exactly, or the data doesn't exist, or is too copious (for example, having to know the position and velocity of every molecule in a gas at a given time), nevertheless very useful *qualitative* results can still be obtained. In fact, by ignoring some information one can end up with more insight (like being able to see the wood for the trees). A remarkable technique that does just this (it throws away data) is the 'surface of section' method discovered by Poincaré in 1892.[24] The astronomers, Hénon and Heiles, put this technique to spectacular use in the 1960s.[25] Even where equations couldn't be solved exactly (analytically), they were able to

[24] Henri Poincaré, *Méthodes Nouvelles de la Mécanique Céleste*, Paris, 1892; Dover Publications, 1957; New Methods of Celestial Mechanics (English translation), edited by D Goroff, American Institute of Physics, New York, 1993.

[25] From G J Sussman and J Wisdom, *Structure and Interpretation of Classical Mechanics*, MIT Press (2001); M Hénon, *Numerical Exploration of Hamiltonian Systems, in Chaotic Behaviour of Deterministic Systems*, North-Holland Publishing Co. (1983).

gain insight by using computers to carry out numerical approximations (but still using Hamiltonian Mechanics). They could tell such things as whether a system was sensitive to starting conditions, whether the solutions would be chaotic or settle down, whether certain regions of phase space were avoided, and so on. Michel Hénon captured the philosophy perfectly when he wrote:

> "Numerical experiments [in Hamiltonian systems] are just what their name implies: *experiments*... [the aim is to] *understand* the fundamental properties of dynamical systems rather than to *prove* them."[26]

7.6 Conservation Theorems

When considering an actual fluid, the metaphor of a phase fluid is especially apt. We can model the motion of a real hydrodynamical fluid in everyday 3-D space in two ways: we can use the 'particle description', or we can use the 'field description'. In the first case we watch the change in position with time of a tiny volume-element ('particle') of fluid; in the second case we look at the 'velocity field' - the instantaneous velocity at each point in the fluid. This three-dimensional hydrodynamical picture carries over completely into the $2n$-dimensional phase space. The remarkable thing is that certain well-known and important conservation theorems apply equally well to real fluids (well, that is to say, idealized zero-viscosity, real fluids) and to the phase fluid. We discuss these now.

1) Liouville's Theorem

   It turns out that the abstract phase fluid has the same properties as a real fluid that is incompressible. An incompressible real fluid is one in which any volume-element of fluid (any sample of neighbouring 'particles') cannot be compressed, that is, it keeps the same volume. Likewise, for the phase fluid we may examine any small $2n$-dimensional volume-element within the fluid, and watch this volume as it is 'carried along' by the flow. Although the shape of the volume-element may become distorted yet its total volume always remains unchanged:

[26] From G J Sussman and J Wisdom, *Structure and Interpretation of Classical Mechanics*, MIT Press (2001); M Hénon and C Heiles, The Applicability of the Third Integral of Motion: Some numerical experiments, *Astronomical Journal* 69 (1964) pp 73–9.

$$\text{volume-element} = \int dq_1, \ldots, dq_n; dp_1, \ldots, dp_n = \text{constant} \quad (7.10)$$

(The case where an integral comes out constant is what the French mathematician, Henri Poincaré (1859–1912), called an 'integral invariant' of the phase fluid.) An equivalent way of stating this, in the 'velocity-field description' of the fluid, is to say that the divergence of the velocity is zero. It would be too much of a digression to explain the meaning of 'divergence' in physics, but the interesting thing to learn is that, for the phase fluid, its divergence is guaranteed to be zero *because* of the canonical equations. This is shown in Appendix A7.4. In other words, the phase fluid is incompressible *by definition.*

2) Helmholtz's Circulation Theorem

Another of Poincaré's 'integral invariants' concerns the 'circulation' of the phase fluid. The 'circulation' is a property of real fluids, and it was discovered by the German physicist Hermann von Helmholtz (1821–94) in 1858. (Helmholtz was also one of the discoverers of 'energy'.[27])

We consider a 'material' line[28] in the fluid - a line that goes from one specific 'particle' to the neighbouring 'particle', and to the next neighbouring 'particle', and so on (a bit like beads on a necklace). Suppose this material line - defined at one given time - forms a closed loop. As time progresses, the loop moves forwards (as it is the line that joins 'particles' which themselves move forwards in time). We can now consider increments, dq_i, along this line, and at each increment form the scalar product with the velocity-field, p_i, at that point (the mass is assumed constant and so the velocity-field is equivalent to the momentum-field). Finally, we can sum up (integrate) all these products. We end up with a closed line integral of the velocity-field around an area-element of the fluid.[29] This is the 'circulation'. In the phase fluid, it is:

[27] Coopersmith J, Energy, the Subtle Concept, Oxford University Press (2015).

[28] (not just a mathematical curve but a line that links specific massy 'particles' together)

[29] Clarification: the summation is over all the different 'particles', $i = 1$ to n; the integral is along the closed loop encircling an area in phase space.

$$\text{Circulation of phase fluid,} \quad \Gamma = \oint \sum_{i=1}^{n} p_i dq_i \qquad (7.11)$$

According to Helmholtz, the 'circulation' of a zero-viscosity hydrodynamical fluid is a constant:

$$\text{Circulation of ideal real fluid} = \Gamma = \text{constant} \qquad (7.12)$$

This has the physical meaning that, given that there is no dissipation (no viscosity), then a fluid which contains no vortices will continue to contain no vortices, whereas a fluid which does contain vortices will continue to contain vortices: in other words, vortices cannot be created or destroyed. The constancy of the circulation in the case of the *phase* fluid likewise means that *vortices cannot be created or destroyed in phase space.*

Commentary 3)
In configuration space, the action integral is over a curve, and the ends of the curve are fixed during variations. When it comes to minimization problems in phase space, however, we have to do with a fluid rather than a curve. Therefore, instead of end-*points* of a curve we now have cross-sectional *areas* through a fluid. Now that we have to do with end-areas rather than end-points, it is appropriate to allow variations at these ends (within the end-area). We therefore relax the usual ban and *do* allow variations of the q_i s at the ends of the action integral.[30] The condition of stationarity in phase space is then:

$$\delta(Action) = \left[\sum_{i=1}^{n} p_i \delta q_i\right]_{t_1}^{t_2} = 0 \qquad \delta q_i(t_1) \neq 0 \text{ and } \delta q_i(t_2) \neq 0$$

The term in square brackets is a 'boundary term' (cf. equation 6.6, Section 6.2,), and, curiously, it has exactly the same form as the circulation. Is this a coincidence? No, nothing in physics happens by coincidence; the boundary term and the circulation are defined in the same way because: both relate to perfect differentials, of dual dimensionality (an area), which exist at a *boundary* (a line going around the edge of the area), and which sum or integrate to a constant (zero, if taken round a closed sum or loop).

[30] We also allow variations of the p_i s at the ends because, as we have said in Commentary 2) and Appendix A7.2, the p_i-variation makes no difference to δ(Action).

Everything we have already remarked about the profound link between conserved properties and symmetries in Lagrangian Mechanics (Section 6.9) applies again in Hamiltonian Mechanics, but more so. As before, conserved properties and symmetries are linked in the following way: a system is 'symmetric' with respect to a coordinate, q_i, if small changes in that coordinate leave everything looking exactly the same. (We are talking here of *continuous* symmetries - not discontinuous symmetries such as 'reflection in a mirror', or 'changing all particles into antiparticles'.) For example, if we have a bench-top experiment and nothing changes after displacing the equipment gently through a small distance, Δx, along the bench-top, then the system displays symmetry in this x-direction. This is the same as saying that nothing in the system depends on x, and so x doesn't occur in the Hamiltonian. We say that x is an absent or 'ignorable' coordinate (cf. Section 6.9).

But (the difference between the two Mechanics is that) Hamilton's Canonical Equations are exceptionally well-suited to the bringing out of conserved properties - because if x is 'ignorable' then it doesn't occur in H, and so $\partial H/\partial x$ will be zero, and then from (7.9) $\dot{p}_x = 0$, and so $p_x = \text{constant}$. Therefore, we have a *conserved quantity arising out of a symmetry*. Specifically, we have a conserved 'conjugate' momentum, p_x, arising from a symmetry with respect to the absent ('ignorable') variable, x. We had similar arguments in the case of Lagrangian Mechanics but now, because of Hamilton's Canonical Equations, the link between symmetries and conserved 'conjugate momenta' is even more transparent. Some well-known conservation theorems that enter in this way are the conservation of linear momentum (arising out of a symmetry with respect to linear translations), and the conservation of angular momentum (arising when the system is symmetric with respect to rotations).

7.7 Hamilton's Mechanics and the conservation of energy

There is another special conservation theorem - the famous conservation of energy - and it arises in two special ways. In the first way, we have a system which does not depend explicitly on time. (There is always an implicit dependence on time, for example, a swing changes its position with time, but it is not explicit - there are no sudden gusts of wind at specific times.) Because t does not occur explicitly in the Hamiltonian, then the phase fluid shows a special kind of motion - steady-state motion. There *is* motion (the velocities are not all zero) but the velocities (the p s) are always constant at their respective q s. For such conservative, steady-state systems, the streamlines in phase space have an additional property - they trace out curves of constant H (we could call them iso-H curves). Actually, they are curves only in the simplest case of two dimensions; generally, in more than two dimensions, the p s and q s 'dance' forever on a hyper-surface of constant H (the subscripts, i, have been left out, for clarity). Now as H, in this special case, does not depend explicitly on t, then we have:

$$\frac{dH}{dt} = \sum_{i=1}^{n} \left(\frac{\partial H}{\partial q_i} \dot{q}_i + \frac{\partial H}{\partial p_i} \dot{p}_i \right) = 0 \tag{7.13}$$

However, the canonical equations tell us that $\dot{q} = \partial H/\partial p_i$ and $\dot{p} = -\partial H/\partial q_i$. Substituting these into (7.13) we obtain $dH/dt = 0$ or, in other words, $H = constant$. We have therefore found that energy is conserved (in this time-independent case). As we remarked already with regards to d'Alembert's Principle (Section 5.5), and Hamilton's Principle (Section 6.7), we have not had to assert energy conservation but have been able to *deduce* it - in the present case by assuming only time-independence and the validity of the canonical equations of motion.

The second way in which a system can be conservative is rather remarkable. We consider the more general case, where there *is* an explicit dependence on time, but then we 'get rid' of t by pretending it's one of the q-coordinates (the number of these thereby increases from n to $n + 1$). Once again, we end up with a fluid flowing in an abstract space, but it is a space of $2n + 1$ dimensions - that is, $n + 1$ q-coordinates and n p-coordinates). The French mathematician, Élie Cartan (1869–1951), called this the 'state space'. In this state space the problem of motion is completely geometrized - it's like taking a photo at every time, and then dispensing with time and looking at an infinite collection of photographs instead. But we can do even better than this. We can absorb t into the position coordinates, as just explained, but then we consider all $2n + 1$ coordinates as functions of yet another coordinate, say, τ. (This makes sense as, having got rid of the independent variable, t, we now need a new independent variable - one which 'puts the photos into order'.) We call this the 'extended phase space'. Then a curious thing happens: having inflated the number of 'position' coordinates, and introduced a new independent variable, it turns out that the new system *is* conservative; t doesn't occur, and to compensate for bringing in the new variable, τ, there is a new condition - the very condition that energy must be conserved. (How this happens mathematically is shown in Appendix A7.5.) In short, any system, even one with an explicit time-dependence, can be recast so as to be conservative.

7.8 The Hamilton-Jacobi Equation

"What I tell you three times is true"
Lewis Carroll, The Hunting of the Snark

This section has much mathematics. The treatment is gentle, but the whole section may be skimmed or skipped without much loss of continuity.

The aim of mechanics is to solve the equations of motion, $q_i = q_i(t)$, for all the generalised particles, $i = 1$ to n, - or is it? Sometimes these equations are impossible to solve (too many, too difficult, or the input data are unavailable) and then we are content to learn other information: general qualitative information about how the system evolves, what are the symmetries, are there any conserved properties? In Hamilton's Mechanics we would like to know how the 'ray property' comes about (how the 'wavefronts' move forwards in time, and why the 'rays' are perpendicular to these wavefronts), what the Hamiltonian is, and why at a given time the ps and qs stay on a surface of constant H. The trouble is, these things happen in *different* spaces (the wavefronts occur in configuration space, whereas the surfaces-of-constant-H occur in phase space). Coordinate transformations will come to the rescue.

It may seem strange that something so arid and mathematical as a coordinate transformation can bring in physical insight but already, at the end of the previous section, we saw how the important energy conservation principle came in this way,[31] and anyway we have already swallowed the pill that a 'mathematical test in an abstract space' is an indispensable, physically-telling procedure. So far in this book we have had three transformations: 1) from everyday space to configuration space; 2) from configuration space to phase space; 3) from phase space

[31] We re-branded t as a position coordinate, and then introduced a new 'dummy' independent coordinate, τ.

to 'extended phase space'. We have already seen the utility of coordinate transformations, but we're about to discover that this is only the beginning. To paraphrase Lewis Carroll (start of section): "What I transform at least three times is still true." We'll need a lot more than three transformations, in fact, we'll need an infinite number of them. It's going to be a long story, involving a little hand waving, and much hard thinking. It goes like this.

A coordinate transformation is something brought about by a function - a function that transforms all the coordinates from old to new, say, from Q_i to q_i, for all i. We want to find that special transformation, call it S, which acts as a 'generating function', the function that makes the 'wavefronts' move forwards in time. What we're going to do is to try and home in on S by subjecting it to more and more stringent conditions.

First, we require that S *does* depend explicitly on time (we drop our usual incantation 'and possibly t') as it will be this very dependence on t that drives the transformation forwards, generating the infinite succession of wavefronts. Next, we stipulate that S must be a *continuous* function: that is, if the input time, t, increases by a small amount, Δt, then the output, q_i, must increase by a small amount, Δq_i, without any gaps. (This is a standard stipulation - most functions we deal with in analytical mechanics are continuous functions.) Another condition we insist upon is that all the transformations must satisfy Hamilton's Canonical Equations. That is to say, we can transform from old coordinates to new coordinates but then the Canonical Equations (7.9) must still apply in these new coordinates. Transformations that comply are known as Canonical Transformations (CTs). Yet another condition that S must satisfy is that the 'circulation' (7.12) must be an invariant (the difference between successive 'circulations' must be zero). This is shown in Appendix A7.6. Curiously, defining S in this way automatically means that it satisfies the following transformation equations (see Appendix A7.7):

$$p_i = \frac{\partial S}{\partial q_i} \quad \text{and} \quad P_i = -\frac{\partial S}{\partial Q_i}, \qquad i = 1 \text{ to } n \tag{7.14}$$

(In fact, even more curious, the three conditions - the transformations (7.14), the invariance of the circulation, and the requirement of being canonical - are all very closely connected.) The special thing about equations (7.14) is that they are reminiscent of the equations for 'grad' (also called del, and symbolized ∇). This is an operator in vector calculus, which in Cartesian coordinates has the form $(\frac{\partial}{\partial x}, \frac{\partial}{\partial y}, \frac{\partial}{\partial z})$, and this is similar to $(\frac{\partial}{\partial q_1}, \frac{\partial}{\partial q_2}, \dots \frac{\partial}{\partial q_n})$ in generalized coordinates. Without going further into the mathematics, we just need to know that 'grad' has the ability to seek out the paths of *steepest* gradient away from a given surface. This is exactly what we need for our programme - it guarantees the 'ray property' because 'steepest' is the same as 'perpendicular' as regards ways of leaving a surface.[32] (The perpendicularity of a path within the wavefront and leaving the wavefront is shown in Appendix A7.9.)

This is very promising, but there is still a problem. S (sometimes called a 'generating function') is a function of q_i and Q_i, and it implicitly defines a transformation between old coordinates, Q_i, and new coordinates, q_i. However, it's a transformation that doesn't cleanly separate the new and old coordinates onto opposite sides of an equation but mixes them up together on both sides of the equation. We would like a transformation that doesn't tangle up forwards and backwards but is purely progressive and, moreover, progressive *in time*.[33] That is, we desire a transformation which expresses new coordinates purely in terms of old coordinates, and all increasing continuously as time increases continuously. This is rather a special desire and it can only be met because of three special factors:

(i) Continuity

At t, S implicitly defines a canonical transformation, CT, that maps some starting position, say, Q_i, into q_i; whereas at the neighbouring (slightly later) time t', S' implicitly defines CT$'$ that transforms the same starting position, Q_i, into q_i' which neighbours q_i (we have $t' = t + \Delta t$ and $q_i' = q_i + \Delta q_i$ where both Δt and Δq_i are small.[34]) In summary:

[32] Relations (7.14) lead to paths that lie along the 'momentum vector', **p**, formed by all the p_i s. By the definition of momentum, **p** has the direction of the tangent to the mechanical path, ensuring the perpendicularity of the 'wavefronts' and the 'rays' as they leave the 'wavefronts'. Likewise, the second set of equations in (7.14) lead to paths that lie along **P** formed by all the P_i s, ensuring the perpendicularity of the 'rays' as they *arrive* at the 'wavefronts'.

[33] As an analogy, consider geneology of the Greek Gods. We can jump backwards and forwards in time and say "Hephaestus was the grandson of the son of the grandfather of Zeus", or, much better, we can say that "Uranus was the father of Cronus who was the father of Zeus who was the father of Hephaestus".

[34] Notation: the prime indicates a transformed coordinate, not differentiation.

At t, S implies CT which does $Q_i \mapsto q_i$
At t', S' implies CT$'$ which does $Q_i \mapsto q_i'$
S' is close to S, t' neighbours t, and q_i' neighbours q_i.

The crucial property of 'neighbourliness' follows from the fact that S and S' are *continuous* functions.

(ii) Canonicity and the group property

As mentioned earlier, S doesn't define just any transformation, it defines specifically a *canonical* transformation. It so happens that canonical transformations satisfy the "group property".[35] This means that if two transformations are canonical then any composition of them will also be canonical. Also, if a transformation is canonical then its inverse will be canonical. Of relevance to our needs, the transformation $q_i \mapsto q_i'$ can be composed of two others: the inverse transformation, $[Q_i \mapsto q_i]^{-1}$, followed by the transformation, $Q_i \mapsto q_i'$. These two are canonical, so by the'group property' we are assured that $q_i \mapsto q_i'$ is also canonical. In short, the transformation $q_i \mapsto q_i'$ may be written as $q_i \mapsto q_i + \Delta q_i$, and it is exactly what we require: it is canonical, the arbitrary starting Q_i has disappeared, and q_i is a 'running variable', moving forwards in time.

(iii) The $p_i \mapsto q_i$ transformation equations, (7.14)

Finally, using (7.14), everything we have just said about transforming the q_i s can be made to apply to the p_i s as well, that is, we can arrive at the canonical transformations $p_i \mapsto p_i + \Delta p_i$. (Once again, the arbitrary starting P_i has disappeared, and once again, p_i is a 'running variable', moving forwards in time.)

Altogether we have:

$$q_i \mapsto q_i + \Delta q_i$$
$$p_i \mapsto p_i + \Delta p_i \qquad i = 1 \text{ to } n \qquad (7.15)$$

These transformations are rather remarkable. They creep forwards in tiny - in the limit, infinitesimal - steps, and each point of phase space is transformed - mapped - into a neighbouring point. As they are infinitesimal and constantly canonical,

[35] If this is unfamiliar, jump straight to the concluding sentence: "In short...forwards in time".

they are known as infinitesimal canonical transformations, ICTs. Remarkably, the transformation itself can change at each instant (through the dependence of S on t), and so we have an infinite number of ICTs generating the motion of the phase fluid in phase space. The outstanding property of an ICT is that it leads to an *explicit* mapping - that is, the starting parameters are on the left, the final parameters are on the right.

That's not all. The q_i and p_i are running coordinates, as desired, but now comes the *coup de grâce*. We've got rid of the arbitrary starting coordinates, Q_i and P_i, but now we'd like to get rid of even our running coordinates and end up with only *relative* coordinates, Δq_i and Δp_i. How we achieve this is we form the difference S'-S, at two close times, Δt, and then allow Δt to become infinitesimal.[36] The mathematics is shown in Appendix A7.8. We finally arrive at:

$$\Delta q_i = -\frac{\partial\left[\frac{\partial S}{\partial t}\right]}{\partial p_i}\Delta t$$

$$\Delta p_i = \frac{\partial\left[\frac{\partial S}{\partial t}\right]}{\partial q_i}\Delta t \qquad i = 1 \text{ to } n \qquad (7.16)$$

The expression in square brackets is $\partial S/\partial t$. As S is an unknown function, then so is '$\partial S/\partial t$' an unknown function, so let's call it X - a good symbol to represent something unknown. (Besides, $\partial S/\partial t$ is messy and too tiny

[36] In physics, even when you can't say something absolute, you can often say something absolute about a *difference*. Also, it is often the case that certain final outcomes can only be achieved by taking *infinitesimal* steps (for example, we have just found that for CTs to be 'progressive' they must be ICTs). A totally different example occurs in the case of rotations. A finite rotation of an object in everyday 3-D space can be made up of a succession of infinitesimal rotations carried out in any order. However, for large rotations, say through 90°, then the order does make a difference. Try it right now! Put your bookmark in, close this book, (or press pause on your e-book) and then carry out the following sequence of rotations: (i) 90° about the spine, then 90° about the base; and then start again from fresh and carry out (ii) 90° about the base, then 90° about the spine. The outcome is totally different in (i) and (ii). However, with a sequence of *infinitesimal* rotations, the outcome will be identical, *irrespective of the order*. We begin to appreciate why the 'absolute geometry', and Variational Mechanics, have to do with infinitesimals, in particular, differentials.

to read.) We can also remove the square brackets, and equations (7.16) become:

$$\Delta q_i = -\frac{\partial X}{\partial p_i}\Delta t$$

$$\Delta p_i = \frac{\partial X}{\partial q_i}\Delta t \qquad i = 1 \text{ to } n \qquad (7.17)$$

These equations have the same virtues as the "remarkable relation" in Appendix A7.8: there are no absolute coordinates, only intervals between coordinates; moreover these intervals are tiny; the time-dependence is neatly together on one side. Finally, we divide both sides by Δt, and allow this time interval to become infinitesimal, and, hey presto, we arrive at:

$$\dot{q}_i = -\frac{\partial X}{\partial p_i}$$

$$\dot{p}_i = \frac{\partial X}{\partial q_i} \qquad i = 1 \text{ to } n \qquad (7.18)$$

These remind us of Hamilton's Canonical Equations, (7.9). In fact, if we say that our mystery function, X, is equal to minus the Hamiltonian function, H, then equations (7.18) *are* the Canonical Equations. In other words, we have discovered that $X = -H$, and therefore that $\partial S/\partial t = X = -H$, or:

$$\partial S/\partial t + H = 0 \qquad (7.19)$$

This is sublime in its simplicity. In addition, we have at long last discovered something about S. (Remember, H is a *known* function - derived from prior knowledge of the given mechanical system, which is subject to given constraints, and a given potential energy function, V.)

We are not done yet. S is a function of q_i, Q_i, and t, and so $\partial S/\partial t$ is, likewise, a function of q_i, Q_i, and t, but H is a function of q_i and p_i (and possibly t) - all for $i = 1$ to n. However in one equation we should stick to one space. We can either convert S into being a function of q_i and p_i, or convert H into being a function of q_i and Q_i. We choose the latter conversion, as we are most interested to find out about how the 'wavefronts' are generated, and this occurs in configuration space.[37] To achieve this we benefit from relations (7.14), and so wherever p_i occurs in H we replace it by $\partial S/\partial q_i$ (this does the trick because S and hence also $\partial S/\partial q_i$ are functions of q_i and Q_i - but never p_i). Thus, we can write out equation (7.19) again, this time with the arguments of H indicated:

Hamilton-Jacobi Equation

$$\partial S/\partial t + H(q_i, \partial S/\partial q_i; ; t) = 0 \tag{7.20}$$

This is a landmark equation, discovered by Jacobi (between 1834 to 1837), and is known as the 'Hamilton-Jacobi Equation'. It is exactly what we want - an equation that links together the Hamiltonian (the energy function that determines the 'dance' between the p_i s and the q_i s) and the 'generating' function, S (that determines how the 'wavefronts' - surfaces of common action - move forwards in time).

Let us pause to look at the geometric and physical meaning of this equation. Which space are the surfaces of common action moving in? As we have explained before, it is configuration space; we know this by comparison with Hamilton's Optical theory (see earlier footnote), and because S depends on the q_i s but not the p_i s. Knowing what space we're in, let's set the clock a-ticking and see what happens. We take the starting S to be a constant, say, zero. The Hamilton-Jacobi equation then determines that as t increases S spreads out, approximately radially, as one forwards-moving surface propagated like a shock wave (cf. the sound wave produced by an explosion). See Figure 7.5.

Why doesn't the wavefront spread out symmetrically (that is, as perfect circles centred on the source)? It is because $E - V$ is usually not uniform over the whole space, so some 'particles' will be speeded up or slowed down relative to other 'particles' (in Section 7.8, we will learn that $\sqrt{2m(E-V)}$, in mechanics, acts as a quasi 'refractive index'). One thing to note is that the surface, despite being called a wavefront, has no *wavy* properties (it is not oscillatory, and has no frequency or wavelength) - this also will be touched on in Section 7.8. Finally, we ask: what of the Hamiltonian? We find that there are surfaces (strictly-speaking, hyper-surfaces) of common H - but these occur in *phase space*.

[37] See the discussion near the beginning of Section 7.3. It was there explained that the wavefronts in optics are surfaces in everyday space, and so the wavefronts in mechanics will be surfaces in configuration space which, exceptionally, reduces to everyday space. (By contrast, phase space never reduces to everyday space.)

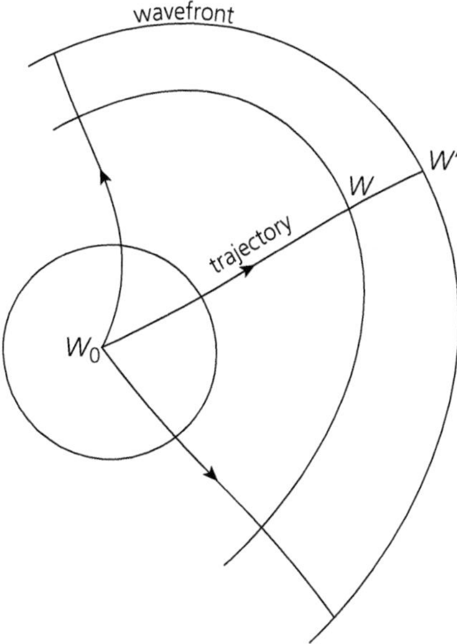

Figure 7.5 Wavefront of common action in configuration space (adapted from Brillouin L, *Tensors in Mechanics and Elasticity*, Academic Press (1964) Fig. VIII.4, page 219.)

It is on these surfaces that the p_i s and q_i s perform their perpetual 'dance'. In time-independent cases, $H = E =$ constant, and we find that there is a linear relation between the surfaces of constant action and the surfaces of constant H ($S = C - Et$, where C is a constant, the surface-of-constant-action at time t).

If we can find out what functional form S has, then we can determine first the $\partial S/\partial q_i$, then the p_i s, and then the trajectories. In other words, knowledge of the function S completely solves the mechanical problem. But this is a bit like saying: if we had the answer, we'd have the answer - so, how can we obtain this function, this 'open sesame' of mechanics? We started off this section saying that mechanics was tantamount to finding the solution to n equations of motion, $q_i = q_i(t)$, each having *one* independent variable, t, whereas now equation (7.20) has brought us to an extraordinary turnaround: our quest is turned inside-out and we desire to solve just *one* equation, with just *one* dependent variable, S, but n independent variables, q_i, (plus also the time, t).

Despite there being only one dependent variable, equation (7.20) is a partial differential equation, and - as we have just admitted - it is not easy to solve (in Hamilton's original version, involving two simultaneous partial differential equations,[38] it is a well nigh impossible problem to solve). In only a very few cases can a complete solution to the Hamilton-Jacobi equation be obtained. We work through two examples in Appendix A7.10. The examples have been adapted from Herbert Goldstein's classic text, "*Classical Mechanics*", and, as Goldstein says, it is a case of "using a sledgehammer to crack a peanut".[39] The trouble is that this method can only be used in very simple problems - just exactly those problems where easier methods are available. The true motivation for using the Hamilton-Jacobi approach is for the physical insights to be learned. The most famous will be discussed in the next section. For now, we end with a brief mention of time.

'Time' and 'energy' are two subtle concepts in physics. We have already seen how in Hamilton's Mechanics the energy function, H, takes centre stage.[40] We are about to learn something more. In the 'extended phase space', where t is re-branded as the $(n+1)$th position coordinate, q_{n+1} (see Section 7.6 and Appendix A7.5), we can ask "what is the conjugate momentum associated with q_{n+1}?" From equations (7.14) we have $p_{n+1} = \partial S/\partial q_{n+1}$, but this may alternatively be written $p_t = \partial S/\partial t$. Now, from the Hamilton-Jacobi equation, we know that $\partial S/\partial t = -H$. Putting the two together, we find that $p_t = -H$. Therefore, saying 'p_t is conjugate to t' is the same as saying '$-H$ is conjugate to t', or, in other words, *'−energy' and 'time' are conjugates of each other.*[41] Confirmation of this startling liaison comes from the fact that energy×time has units of *action*, but we can push our understanding even further. Not only does 'time' often act as a 'running coordinate', but a 'running coordinate' stripped bare of all other duties. It is then hardly surprising that its conjugate coordinate has structure and extensivity galore - it *is* the energy.

[38] Lanczos, page 227, equations (79.20).

[39] Goldstein H, *Classical Mechanics*, 2nd Edition, Addison-Wesley, 1980. Goldstein makes this comparison when solving a simple problem with canonical transformation theory.

[40] (see the comments about H in Section 7.5)

[41] Furthermore, when t is 'ignorable' (the system doesn't depend on time) then p_t, the energy, is constant - as expected.

7.9 The royal road to quantum mechanics[42]

"Put off thy shoes from off thy feet, for the place whereon thou standest is holy ground." [43]

Old Testament, EXODUS III, 5

From Hamilton's original insight of the fundamental identity between optics and mechanics came eventually (a hundred years later) the wave-particle duality, and quantum mechanics. (There were, of course, new experimental phenomena and additional physics postulates that were required for this next development.) Let us follow how each theory (optics and mechanics) helped the other one along, and how quantum mechanics emerged.

Some 150 years before Hamilton, during the time when the corpuscular theory of light held sway, the Dutch diplomat and natural philosopher, Christiaan Huygens (we have met him in Chapter 2), put forward a *wave* theory of light. He proposed, in (1678), in the Huygens Principle, that every point on a surface-of-simultaneous-arrival of light acts as a secondary source of light. These infinite number of adjacent sources emit the light (forwards[44]) in all directions. At the next instant of time, the new surface-of-simultaneous-arrival - or '*wavefront*' - is the envelope of all these secondary wavelets. From this new wavefront there will be another infinity of light-sources, producing another infinity of wavelets, and the next wavefront may be constructed - and so on, and so on (see Figure 7.6).

From Section 7.8 we see that Huygens' Principle is also the basis for constructing the wavefronts-of-common-action that occur in Hamilton's Mechanics. Staying first with optics, we zoom in and scatter little hypothetical 'porcupine quills' over a wavefront, each one perpendicular to the surface, and each representing the path of a light ray. Suppose we wish to find out what will be the new surface after a small interval of time, say, ϵ^{time}. How we find this out is we mark off the small distance, $\Delta(light)$, travelled by light along a quill during the

[42] The quote has been adapted from "A royal road to quantization", Arnold Sommerfeld, as in Goldstein H, *Classical Mechanics*, page 483.

[43] As quoted in Lanczos, page 229. Regardless of religious beliefs, one is moved by the cultural freight of these words, written around two thousand five hundred years ago.

[44] The Principle does not explain why the wavefronts aren't also generated to move backwards.

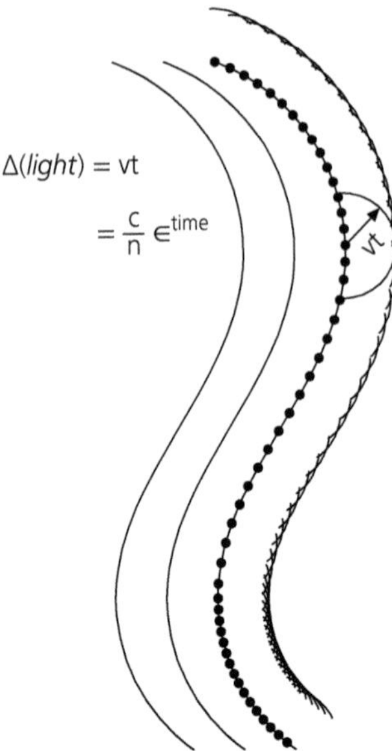

Figure 7.6 Huygens's wavelets.

time ϵ^{time}. We do the same thing for many quills (rays), always starting from the same starting wavefront, and always for the same time interval, ϵ^{time}. The magnitude of $\Delta(light)$ along a given quill is evidently governed by the local speed of light, itself determined by the refractive index, $n(x, y, z)$, of the optical medium in the vicinity of that quill. Finally, we join up all the ray-tips, and and together they define the new wavefront-surface.

It's a similar procedure in mechanics. We have some agreed starting wavefront, W^{start}, and then we scatter little hypothetical 'porcupine quills' over it, each quill *perpendicular* to the wavefront, and each one representing a mechanical 'ray'. We want to find out what will be the new wavefront after a small interval of action, ϵ^{action}, and so we mark off the small distance, $\Delta(particle)$, travelled along the quill, starting from W^{start}, 'during' ϵ^{action}. Now our interval, ϵ^{action}, is the *same* whatever ray (quill) we're following, but the increment travelled along the ray (the magnitude of $\Delta(particle)$) will be different for each ray, governed by the 'refractive index' in that local vicinity. Finally, we join up the tips of the rays, and together they define the new wavefront, $W^{new} = W^{start} + \epsilon^{action}$.

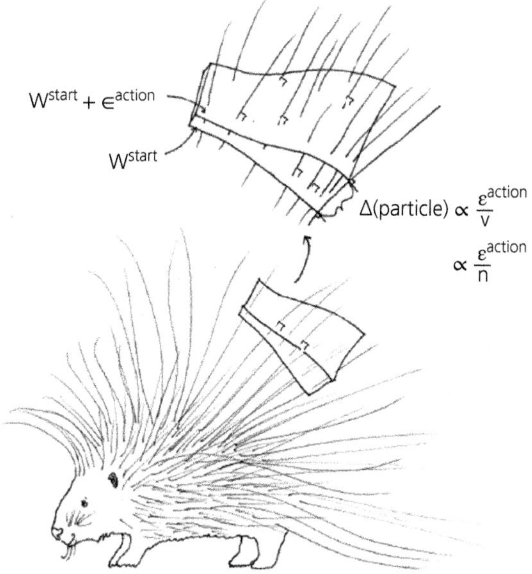

Figure 7.7 Construction of the next wavefront.

These wavefront constructions are governed, ultimately, by equations. We know that in *mechanics* we have the Hamilton-Jacobi equation, but now we ask - is there an equivalent (first-order, partial differential) equation that generates the surfaces in *optics*? Yes, there is such an equation in geometrical optics, it is called the Eikonal Equation (see Appendix A7.11). This Appendix is optional reading, but we transport a result back into the main text. It is that in mechanics, as we have already hinted, we may imagine that the 'rays' pass through a hypothetical medium with a certain 'refractive index': we can now state that (in conservative cases) this hypothetical 'refractive index' is proportional to $\sqrt{2m(E-V)}$.

A historically startling fact: It is humbling to learn that over three hundred years ago the Swiss mathematician, Johann Bernoulli (him of the *brachystochrone* curve, and the Principle of Virtual Work, see Chapter 2), treated the motion of a particle of constant mass m falling under gravity as if the particle was a quasi light-ray moving through a fictitious medium with a gently graduated refractive index. He took the refractive index to be proportional to $\sqrt{(E-V)}$.

Despite this close analogy between optics and mechanics, it is not perfect - after all, light and massy particles are not the same. The analogy holds good when it concerns the *paths*, but not when it concerns *how fast* the particles travel along those paths. We know that, in the usual way, the distance, $\Delta(light)$, travelled by light during a given time-interval, ϵ^{time}, is proportional to the speed of light; however, in mechanics, the 'distance', $\Delta(particle)$, travelled along a 'ray' during a given ϵ^{action} is *inversely* proportional to the speed of that particle. (So, in mechanics, the wavefronts are closer together the faster the particles, while in optics the wavefronts are further apart the faster the light.) There are some other caveats which we give below as optional extra information.

Optional Commentary

Caveats of the 'ray property'

(1) Whether in optics or mechanics, we preselect for consideration rays that already *are* perpendicular to the wavefront.
(2) We preselect for our consideration only rays that are monochromatic (optics), or monoenergetic (mechanics).
(3) We preselect rays that are perpendicular, yes, but not necessarily in the *Euclidean* sense. (For example, for most crystals, the light rays are not usually orthogonal to the surfaces; and an electron moving in a magnetic field does not cross the surfaces $S = constant$ perpendicularly.) However, the orthogonality is restored if we think in terms of 'intrinsic' rather than Euclidean geometry.
(4) The optics 'field' is static (the shape of the lenses and mirrors, and their refractive indices, are fixed), but this isn't generally true for mechanics.

We have an even bigger caveat than all those above - it is that the whole of *geometrical* optics is only an approximation to the truth. We learn the laws of geometrical optics in high school (the laws of reflection and refraction of rays, how to draw little diagrams showing upside-down trees in order to determine focal length, and so on). What is often not emphasized is that all these laws are only approximate. They only apply when the angular spread of rays in a beam is 'small', and when the wavelengths are infinitesimal. (This makes sense - it is exactly in the case when the wavelengths are insignificant that ray-like behaviour predominates.)

So, we have rays, and we have wavefronts, but what about the actual waviness of waves - properties like periodicity, frequency, wavelength, and superposition? All these attributes come in when we go over into the theory known as *physical* optics.[45]

In physical optics we assume that the optical disturbance is not merely a ray but a wave - having an oscillatory (up and down) amplitude, and having a definite frequency, ν. The wavefronts then become surfaces of equal *phase* (for example, we might watch the very wave*crest* as it moves along). Therefore, instead of having surfaces-of-simultaneous-arrival-time, say, ϕ, we now have surfaces-of-given-phase, say, $\bar{\phi}$, such that $\bar{\phi} = \nu\phi$. Now, from the theory of geometric optics, there is an equation for ϕ (see Appendix A7.11), and we learn that $|\nabla\phi|^2 = 1/(speed)^2$, and so this will become: $|\nabla\bar{\phi}|^2 = \nu^2|\nabla\phi|^2 = \nu^2/(speed)^2$. Here, (*speed*) is the speed of the wave-surface, but we also know (from the usual theory of waves) that $(speed) = \nu\lambda$, where λ is the wavelength of the wave. Putting this altogether, we find that: $|\nabla\bar{\phi}|^2 = 1/\lambda^2$.

Everything in the previous paragraph relates to optics. Let us now employ Hamilton's optico-mechanical analogy and shift the discussion into the realm of mechanics. Specifically, we consider the case of an electron in orbit around the atomic nucleus. An electron is a particle, but we will forget about this, for the moment, and consider only its path around the nucleus. Moreover, we'll consider its path as a kind of 'light ray'.

The path is a *closed* orbit, and therefore, in order not to have destructive interference of the 'light ray', the orbit must be made up from a whole number of wavelengths. This is equivalent to saying that the phase change between the beginning and end of the orbit must be proportional to a whole number, $\Delta\bar{\phi} = 1, 2, 3, \ldots$ We can word this another way: the phase change, $\Delta\bar{\phi}$, can only take on certain *discrete* allowed values. At this stage of the history of physics, various experimental results to do with atoms came into the theory:[46] it was found that in the atomic regime certain physical features were always found to have tiny *discrete* values, or 'quanta' (the physical features were said to be quantized). There were Bohr's quantized energy levels, and Sommerfeld's and

[45] (Physical optics is a more correct theory than geometrical optics, but it is still not the last word - for that we need Maxwell's equations, and we need to allow for the possibility of vector fields.)

[46] (Of course, one cannot arrive at quantum mechanics wholly from Hamilton's Mechanics.)

Wilson's quantized orbit-areas in phase space. Then Einstein came into the story, with his unearthly ability to get to the heart of a problem. Sommerfeld's and Wilson's results depended on having a special choice of coordinate system, but Einstein saw that the truly fundamental parameter was a sum over the total *action*, S, and using this knowledge he could remove the dependence on an arbitrary choice of coordinates, and arrive at an invariant outcome. (How this comes about is explained in Appendix A7.12. Strange to say, but Einstein's vital contribution to the transition from Hamilton to Schrödinger has been almost entirely forgotten.) This is the same S as occurs in the Hamilton-Jacobi Equation, and it is proportional to the phase, $S = h\bar{\phi}$, where h is a constant of proportionality. In short, the condition $\Delta\bar{\phi} = 1, 2, 3, \ldots$ in 'electron-optics' becomes $\Delta S = h, 2h, 3h, \ldots$ in 'atom-mechanics', and also, the relation $|\nabla\bar{\phi}|^2 = \frac{1}{\lambda^2}$ becomes $|\nabla S|^2 = \frac{h^2}{\lambda^2}$.[47] But we also know that $|\nabla S|^2 = 2m(E - V)$ (we showed this earlier in Appendix A7.11). Furthermore, in the case where $E = T + V$ then $(E - V) = T = \frac{1}{2}mv^2$ (where v is the speed of the electron and m is its mass). Putting this altogether we have: $|\nabla S|^2 = m^2v^2 = \frac{h^2}{\lambda^2}$. Taking the (positive) square root, and rearranging, we finally arrive at a famous result:

$$\text{'de Broglie wavelength'}, \qquad \lambda = \frac{h}{mv} \qquad (7.21)$$

What this equation states is that a *particle* of momentum, mv, may be considered as a *wave*, with wavelength, λ, - the celebrated wave-particle duality. This result was postulated by the French physicist, Louis de Broglie, in 1923 (and published in 1924), after an inspired merging of theory and experimental findings.[48] The postulate applies to all particles (not just to electrons), and also applies in reverse - any radiation will have a corresponding momentum. Why do we not associate a billiard ball with a corresponding wave? Ah, it's all to do with that constant of proportionality, h. We mentioned it earlier, in passing, but never said what it's magnitude was. It turns out that h is more than just a 'constant of proportionality', it is Planck's Constant,[49] a universal constant,

[47] Note that Δ means 'a small increment', and ∇ (called del, or grad) is an operator in vector calculus.

[48] The main experimental result was Einstein's photoelectric effect in 1905.

[49] Max Planck, (1858–1947), discovered (somewhat reluctantly) that the radiation emitted by a 'black body' was quantized, $E = h\nu$. (By the way, it wasn't Planck who attached his name to h.)

having a magnitude that can be experimentally determined. It is: $h = 6.62607004 \times 10^{-34}$ Js. This very very small magnitude means that the wavelengths in (7.21) are very very small - especially in comparison with the typical '*mv*' for a billiard ball - and that's why a billiard ball doesn't do wavy things.

We have admitted that geometrical optics applies only in the case of vanishingly small wavelengths. However, it is the low-λ limit of a more accurate and general theory - one which applies also to finite λ. The more general and accurate theory is a theory in physical optics, and it uses a 'wave equation' discovered by the French physicist Augustin-Jean Fresnel (1788–1827).[50] Now the pertinent question is: in mechanics, might it be that the Hamilton-Jacobi equation is, likewise, the low-λ limit of a more accurate theory? That is to say, is there another 'wave equation', the mechanical counterpart of the Fresnel equation, and which accounts for certain mechanical phenomena when λ is finite?[51] Yes, there is such a 'wave equation' in mechanics - it is known as the Schrödinger Equation.[52] *This is it*, this is the point where quantum mechanics throws a rope bridge to classical mechanics, across the chasm.

The Schrödinger Equation is a famous equation, the foundation of 'wave mechanics' also known as 'quantum mechanics'. (This is the same Schrödinger who we quoted in Section 7.4, praising Hamilton, and 'the Hamiltonian'.) Now Fresnel's Equation is an equation in optics, and it is a function of the *phase* of the wave, ϕ, whereas Schrödinger put forward a (time-independent) wave equation dependent on the *amplitude* of the wave. He called this amplitude ψ.

There is the intriguing question: how does Schrödinger's ψ function tell us where the particle is - in other words, what is it that's waving? When it comes to light, it is the intensity (the amplitude squared) of the oscillating electric and magnetic fields at position, x, which determines the intensity of the light at x (we simplify the discussion by considering just one space dimension, x). A clue, in quantum mechanics, comes from a curious fact: the speed of a particle is *not* the same as the speed

[50] The wave equation is: $\frac{\partial^2\phi}{\partial x^2} + \frac{\partial^2\phi}{\partial y^2} + \frac{\partial^2\phi}{\partial z^2} - \frac{n^2}{c^2}\frac{\partial^2\phi}{\partial t^2} = 0$. To honour Fresnel, Lanczos refers to this as 'Fresnel's Equation'.

[51] (although λ would still have to be 'small' - crucially, of the same order of magnitude as '*mv*')

[52] Erwin Schrödinger (1887–1961), Equation: $\frac{\partial^2\psi}{\partial x^2} + \frac{\partial^2\psi}{\partial y^2} + \frac{\partial^2\psi}{\partial z^2} + \frac{8m\pi^2}{h^2}(E-V)\psi = 0$.

of its wave.[53] Now it is well known that the speed of an infinitely long wave (a perfect sin or cos wave) is given by $\lambda\nu$ where λ is the wavelength and ν is the frequency. However, a perfect wave, defined by just one ν, never exists in nature (just like we never have a perfect point or a perfect line). In practice we always have a small spread of ν s, and then summing over the amplitude-contributions from all these ν s we can end up with a wave that is not infinitely long but forms a localized bump or 'wave-packet', $\psi(x)$, centred at x. (It is essential that ψ has the property of superposition, that is, it can be added, and then the different contributions can combine constructively or destructively.) It turns out that the speed of this wave-packet *is* identical to the particle's speed. This suggests that maybe the wave-packet, taken as a whole (ignoring internal details), is the particle? Yes - and it also turns out that the intensity of the wave-packet, $|\psi(x)|^2$, tells us the *probability* of finding the particle at the position, x. We shall have more to say about quantum mechanics in the next chapter, but it must be admitted that everything in this last paragraph arises from purely quantum effects rather than from an extension of Hamilton's Mechanics.

[53] (We have met this recently, soon after Figure 7.7 - the speed of the wavefronts, S, is not the same as the speed of the 'particles' along the perpendicular 'rays'.)

8

The whole of physics

We give a broad survey, at a semi-popular and qualitative level, of how the Principle of Least Action and the Variational Mechanics come into all of physics. One everyday example occurs in a modern wonder - the smartphone. Instead of measuring positions and angles in the old way using sightlines (relative to a presumed 'absolute space'), the phone contains three tiny accelerometers and measures tiny motions in three directions of the local gravity-field - a testimony to the approach taken in Variational Mechanics.

8.1 Classical mechanics

Before we can analyse a given problem we must first identify what are the 'particles', and what are the independent 'motions' they undergo. Then, we must supply the functional forms of the kinetic and potential energies. After this homework, an amazing simplicity follows. The system follows that path through time, between prescribed initial and final states (never very far apart), that minimizes the difference between the kinetic and potential energies - uses up the least action. For example, the rotating swing-seats at the funfair finely adjust their height so that they achieve exactly the right balance between kinetic and potential energy, at each instant. Also, a ball thrown vertically up against gravity adjusts its speed and height, instant by instant, so that again exactly the right balance is struck between an increasing potential energy and a decreasing kinetic energy as it rises, and between a decreasing potential energy and an increasing kinetic energy as it falls.

For free 'particles' (there are no *external* forces), it is possible to consider the constraints as introducing a 'curvature' into 'space'. If the constraints are static (unchanging through time) then the 'particles' follow the 'straightest lines' - the geodesics - of this 'curved space'.

We have said many times that the Principle of Least Action cannot deal with dissipative effects. However, this is entirely a question of the

The Lazy Universe. Jennifer Coopersmith, Oxford University Press (2017).
 DOI 10.1093/acprof:oso/9780198743040.001.0001

formalism: if the effect can be expressed in the form of a function, then it can be incorporated into Variational Mechanics. (For example, Lord Rayleigh, in 1873, proposed a functional form for velocity-dependent frictional forces.) Whenever such a dissipation function is used, it has the effect of bringing a 'right-hand side' into the Lagrange Equations (see Appendix A6.5).

8.2 Light and electromagnetic waves

It is one of the founding tenets of Special Relativity that the speed of light (in a vacuum) is a constant - but let us change the emphasis to throw relevance towards the Principle of Least Action: light is the *limiting* speed at which any signal can be transmitted. It is hardly surprising, therefore, that light will exhibit extremal behaviour (cf. the Principle of Least Time, which is a version of the Principle of Least Action, and which leads to the laws of reflection and refraction of light).

Consider electromagnetic waves (light is an example). We have an electrical oscillation which generates a magnetic oscillation which generates an electric oscillation and so on and so on (in accordance with Maxwell's Equations). It would not do for the electric and magnetic oscillations to get out of step - they could generate each other to infinity or cancel each other out to zero, both very unphysical outcomes. The extent to which each generates the other depends on the rate of change of each, and so, in order not to get out of step, the electric oscillations and the magnetic oscillations must travel at the same speed and just the right speed - the speed of light, of course. Another equivalent way of saying this is that the total action must be minimized.

8.3 Special Relativity (SR)

SR requires that the speed of light (c) is a universal constant, but as c is still just a speed then it is still defined in the usual 'speedy' way as 'distance over time'. However, as c is an absolute, then it is evident that distances and times must now relinquish this status. The consequences are profound: 'space' becomes four-dimensional - it's known as space-time; moving clocks run slow; moving rulers may shrink; and, most amazing of all, energy and mass are equivalent, $E = mc^2$.

Something that *is* invariant (the same for all observers) is the 'distance interval', *ds*,[1] between 'events'. An 'event' is a location in spacetime, often selected to mark a special happening such as a birth, an explosion, or the start and end of a given particle's worldline. *ds* only manages to be invariant by mixing up space and time in the following way: $(ds)^2 = [(\Delta position)^2 - (\Delta time)^2]$. (This is in units where $c = 1$.) From this defining equation, we find that (other things being equal) *ds* is smaller as $(\Delta position)$ is smaller, and smaller as $(\Delta time)$ is larger. Here comes the clever bit: by considering everything from the point of view of the *rest* frame, *rf*, then $\Delta(position)^{rf}$ is definitionally zero, and so $ds \equiv \Delta(time)^{rf}$, and furthermore $\Delta(time)^{rf}$ is automatically a maximum (non-moving clocks run fastest of all). (A 'clock' could be an excited caesium atom, a person, a decaying muon, and so on.) Time in the rest frame is called proper time and symbolized, τ, and so with this notation we obtain $ds \equiv \Delta(time)^{rf} \equiv \Delta\tau$. Also, we pass from the finite interval, $\Delta\tau$, to its differential version, $d\tau$.

We now inject a result from Special Relativity. It is that the Lagrangian for a free uncharged particle of rest mass m_0 has the form $L = -m_0c^2 d\tau/dt$. (As mentioned at the end of Chapter 7, we can't answer to all physics from the Principle of Least Action alone). This means that the action integral becomes $A = \int -m_0c^2 d\tau$, taken between events at τ_{start} and τ_{end}. Notice an important fact: because of the minus sign in the Lagrangian then minimizing action, *A*, is equivalent to maximizing proper time, τ (the multiplying factor, m_0c^2, is a constant). In short, the Principle of Least Action has been translated into the Principle of Maximum Ageing.

An objection: by choosing the rest frame, haven't we gone against the spirit of the Principle of Relativity and selected a special reference frame? In a sense, yes, but it's an unambiguous choice, and different observers can still determine and agree upon the elapsed time that *would* be displayed on a clock in their own frame of reference. Even more important, by the Principle of Relativity, the outcomes are transportable and apply in all other (valid) reference frames. This is consonant with our experience: we find that the world-line with the maximal ageing *is* the

[1] This is the 'metric' or Riemann distance function, *ds*, that we mentioned in Section 3.6, Chapter 3.

one chosen in nature - things age, they do not 'youthify'.[2] This everyday observation comes from a totally different arena of physics - the Second Law of Thermodynamics. Now as the Second Law is an empirical law, it doesn't prove the case - but it is hard to see how the Second Law could ever have emerged from a Principle of *Least* Ageing. One other objection: how can we go from a minimization problem to a maximization problem? The 'trick' is explained in an analogy - the twins' conundrum. Jack wants to prove to Jill that his bicycle is the slowest - but how can anyone ride as slowly as possible? Jill comes up with a solution: knowing that one can't ride slowest, they must swap bikes and each try and ride the fastest.

8.4 Electrodynamics

Everything in 8.3 applies again, and this is hardly surprising as Einstein's landmark 1905 paper introducing Special Relativity was entitled "*Electrodynamics of moving bodies*". We mention electrodynamics here mainly to note that it is the first arena in which the potential energy, V, *does* depend on velocity (the velocity of an electric charge in a magnetic field), and, by the way, it is thereby the first arena in which Newtonian Mechanics fails. The kinetic energy, T, must also be defined in a new generalized way (the mass is no longer a constant, and the form is no longer purely 'quadratic' in the speeds). The Principle of Least Action carries on as usual - it is only necessary to ensure that the Lagrangian is of the appropriate 'covariant' form.

Finally, there is in electrodynamics an empirical law (really a 'rule of thumb'), known as Lenz's Law, which finds its true explanation in the Principle of Least Action. Lenz's Law states that the direction of an induced current is always such as to oppose the change which produced it. For example: drop a permanently magnetized iron pellet down the shaft of an upright piece of copper pipe, and you will find that the descent of the magnet is noticeably slower than when dropped outside the pipe. The explanation is that the electric currents induced in the pipe have associated magnetic fields which retard the magnet's motion. Yes, but the bigger explanation is that these various microscopic responses happen in just the right way such as to guarantee the Principle of Least Action.

[2] (There is no word because this never happens.)

8.5 General Relativity, Einstein's Theory of Gravitation (GR)

The spacetime of SR is 'flat' whereas in GR there are large energy-sources, and large momenta, and these distort spacetime, making it 'curved'. The Principle of Maximum Ageing applies as before, even in the vicinity of a star, neutron star, or black hole: that is, a freely-falling body follows the path of maximum proper time (between given events), even in this curved spacetime. The result is amazing - gravity is totally explained by geometry, *without* need of a force.

One difference from SR is that the proper time is now not only influenced by motion ('moving clocks run slow') but also by sources of gravity ('clocks near a very large mass run slow'). Consider a wrist-watch in orbit around Earth. If it is displaced to a slightly smaller radius its orbital speed will be greater but time will be more dilated whereas for a slightly larger radius its orbital speed will be slower but time will be less dilated. The time dilation (from both SR and GR) and the orbital speed act in opposition. The actual orbit is the one along which wristwatch-time is stationary - maximized.

We have stressed again and again that the start- and end-states in the action integral must be 'close together', but what if one wants to examine vast tracts of 'space' (in other words, what if one wants to arrive at a global theory)? The remedy in Einstein's Gravitation Theory is to cover 'space' with a patchwork of small regions, and then stitch them together, with no 'puckering' at the joins. The more strongly the spacetime is curved, the smaller the local reference frames must be. Thus, instead of one reference frame of infinite extent and Newton's Law of Gravity, we have a multitude of reference frames of only local extent, and the Principle of Maximum Ageing. This is more philosophically correct, and means that even strongly curved regions can be explored (where Newton's Mechanics would break down).

Einstein's Theory of Gravitation (and Variational Mechanics) is yet more remarkable: it is a complete theory, that is, it not only tells how bodies will move due to the curvature of spacetime, but also how that curvature is *generated*. Instead of one vectorial 'force of gravity' there are many scalar contributions at every point in a 'field'. These contributions are summarized neatly in a mathematical object called a tensor, known as the 'stress energy momentum' tensor. Why not just mass being the source of gravity? It is because of Einstein's mass-energy equivalence (so mass alone doesn't cover all possibilities), which in turn arose out of the need for 'Relativity' between observers. The sources of gravity are: energy density, momentum density, and stress (a bit like pressure). Strange to say, but it's the same in materials science - there is a stress energy tensor which determines how a slab of some material may shear, bend, twist, or stretch (all deviations from the Newtonian assumption of rigidity). One can even say that materials science has a family connection with Einstein's Gravitation Theory.

8.6 Hydrodynamics

In view of the parallels between real fluids and the phase fluid (Section 7.5) it is hardly surprising that the methods of variational mechanics can be applied in hydrodynamics and hydrostatics. However, the fluids (which could include gases) must be 'ideal', that is, they must not have any viscosity. This is because viscosity would introduce frictional effects, and these cannot easily be treated by variational methods.

The advantage of using variational mechanics is that it is not necessary to know the forces that maintain certain kinematical conditions. For example, there may be strong forces that prevent the molecules of a fluid from being squashed close together, and weak forces resisting a change in shape of a given volume-element of fluid. We don't need to know either of these forces, we can simply assert (have as our 'kinematic condition') that the volume of the fluid remains constant.

8.7 Statistical mechanics

In classical mechanics we can usually use the Principle of Least Action directly because we have perfect information about the system (apart from knowing the solutions, of course). In statistical mechanics, however, we have very many degrees of freedom, and we do not have perfect information. For example, we may wish to determine the velocity of molecules in a gas at some time in the future, but as there are trillions upon trillions of molecules (for example, around 10^{25} in a glass of water including ice cubes) then it is impossible to know the positions and velocities of every one of these molecules, either now or at any other time. An entirely different strategy must be adopted. The phase space is perfect for the job, and it was the American physicist and physical chemist, Willard Gibbs (1839–1903), who realized this, and who indeed coined the term, phase space, and founded the branch of physics known as statistical mechanics.

In statistical mechanics, we give up trying to follow the precise changes of state in one system, and look instead at the range of possibilities in a very large collection ('ensemble') of hypothetical identical systems (identical apart from the question of what state they're in). The distribution of states is plotted in phase space, and for each distinct possible state of the system there is a unique point of the space. Finally, from an examination of the behaviour of the whole ensemble, we can make predictions of what may be expected, on average, for the one system of interest.

Note that, in hydrodynamics, statistical mechanics, physical chemistry, and so on, there are so very many degrees of freedom (for example, around 10^{27} air molecules in a typical room) that Hamilton's doubling of coordinates (say, from 10^{27} to 2×10^{27}) is no big disadvantage.

8.8 The quantum world

In common with statistical mechanics, in the quantum world we are always dealing with imperfect information. In fact, despite the incredible success of our theories (for example, the precision and agreement with experiment of quantum electrodynamics is outstanding and without parallel in the whole of physics), hardly any of the problems are capable of exact solution. Fortunately, the Hamiltonian often differs only slightly from the Hamiltonian for the rigorously solvable problem, and then it becomes possible to use certain well-tried methods of approximation, such as 'canonical perturbation theory'.

But let's step back and ask - what is quantum mechanics? Suppose we have the physical laws describing billiard-ball collisions, and then we scale down the billiard balls to $\frac{3}{4}$-size and use a $\frac{3}{4}$-size billiards table and a $\frac{3}{4}$-size cue - we should still expect those same laws to apply. What if we scale down by 0.1? 0.01? 0.001? Will the same laws still apply? Of course, we shall have to attend more and more carefully to things like air-resistance, and the roughness of the baize, and so on, but we have the feeling that we should be able to keep scaling down - after all, the laws have 'm' for mass, they do not specify a mass range of, say, 0.15-0.17 kg. We can make things smaller, and smaller, but eventually we do arrive at a different regime, a regime where the old laws don't apply, and where a new 'mechanics' must be used. We could say that quantum mechanics defines that regime where things are not just smaller, they are *absolutely small*. The next question is: how small is absolutely small?

Ah, we know this from where Hamilton's optical analogy goes over into Schrödinger's Wave Mechanics. This occurs when the dimensions in the problem are smaller than, or of the order of, that constant of proportionality, h. (Looking back at Chapter 7, we can remind ourselves that h is the proportionality between the action and phase surfaces, S and ϕ; and the constant of proportionality between λ and $1/p$ in de Broglie's 'wave-particle duality'; and also h is the proportionality between the energy, E, and the radiation frequency, ν (as found in the relation, $E = h\nu$).) As we mentioned in Chapter 7, h is tiny (it has a value of $6.62607004 \times 10^{-34}$ Js), and so this is where 'absolutely small' begins. But let us notice another thing about this crucial arbiter of smallness - *its units are the units of action.*

'Action' comes into the quantum world in so many ways (and remember that, from Hamilton's optical analogy, we have learned that quantum mechanics is the true accurate theory, to which classical mechanics is a mere approximation.) Apart from those already noted, it gives the order of magnitude of the discrete chunks - known as 'quanta'. For example, the energy of the radiation emitted by a 'black body' (the Planck relation mentioned above), and the allowed energies for an electron in an atom. (We are familiar with these ideas now, but they were shocking and revolutionary at the beginning - a bit like saying someone is only allowed to run at certain speeds.) The action, in the form $h/2\pi$, also occurs in the Heisenberg Uncertainty Relationship, and here it sets the limit of precision to which the conjugate coordinates (momentum and position; or energy and time) can be determined *simultaneously* for *one* particle. It also occurs in the Dirac commutator relations (discovered by the British theoretical physicist, Paul Dirac (1902–84)) which show that the order in which measurements of these conjugate coordinates are carried out does make a difference - a tiny difference of magnitude $h/2\pi$.

Never again, post quantum mechanics, can there be any doubt about the deep significance of action in physics. Moreover, there is a version of quantum mechanics (strictly speaking, quantum electrodynamics) which uses the Principle of Least Action directly, albeit with a quantum-mechanical twist. This is the 'many paths' theory of Feynman.[3] In this theory, the quantum particle (be it a photon, a positron, an electron, and so on) goes from an initial event to a (nearby) later event along

[3] Feynman R P, and Hibbs A R, *Quantum Mechanical Path Integrals*, McGraw-Hill, 1965; Feynman's brilliant popular account, *QED: The Strange Theory of Light and Matter*, Princeton University Press 1985.

each and every path - there's an infinity of them (shades of Huygens Principle). The final amplitude is found by adding up the contributions from all these alternative paths. However, as the different paths have different lengths, then the phase-advance will be different and not necessarily equal to a whole number of wavelengths, and so the waves will not arrive in step. Only a very few nearby paths will add coherently and contribute significantly to the final amplitude; most paths will add destructively and yield no net contribution. (As Freeman Dyson reported it:[4] "Dick Feynman told me... "The electron does anything it likes... It just goes in any direction at any speed... however it likes, and then you add up the amplitude and it gives you the wave-function." I said to him, "You're crazy." But he wasn't.") Feynman's beautifully simple theory not only agrees very precisely with experiment but explains away the teleological objection - that the particle doesn't 'know' which path to follow - wrong, it follows *all* paths. Also, the theory goes over into the classical limit: for massive particles, say, a football moving at 10 ms^{-1}, the wavelength is so miniscule (around 10^{-34} m) that there isn't the remotest hope for paths to interfere constructively - only one path survives, the very path predicted by the Principle of Least Action.

4 Reference 13 in Edwin F Taylor's "A call to action"; see Taylor's excellent website, www.eftaylor.com.

9

Final words

"We have done considerable mountain climbing. Now we are in the rarified atmosphere of theories of excessive beauty..."

Cornelius Lanczos[1]

The Principle of Least Action postulates that when a physical system goes from a prescribed initial state to a nearby prescribed later state, the path connecting these states is a path of tiny incremental changes in action, and, of all possible paths, the path actually followed is the one for which the total change in action has exactly the same value as it does for all infinitesimally close-by paths - in other words, the actual path occurs in a 'flat' region of the 'space-of-paths'. This is a mouthful, much more so than '$\mathbf{F} = m\mathbf{a}$', but if we ever do discover a 'TOE' (Theory of Everything), expressible as one long equation, then you can be sure that every term in that equation will require a whole library of books to explain it.

Granted that we have to do with a minimum principle, there are two main inputs into the Principle of Least Action postulate. There is the physics input - that what we are minimizing is in fact *action*; and there is the mathematics input - that determines what in fact is the 'shortest path' for a given 'space'. There is even a third input - the requirement of philosophical fitness. We discuss these inputs in turn.

The physics input. This asserts that it is 'action' which is the primal physical quantity. This is plausible because action is connected with the all-important energy functions (the kinetic energy, and the potential energy, V); it is the telltale of absolute smallness (where the quantum world begins); and it is made out of the two conjugate variables, p and q. That these conjugate coordinates are the definitive ones is demonstrated in countless scenarios. For example, take the cases of: (i) a gas expanding to fill its container, (ii) the clumping of cosmic matter into stars and planets - do (i) and (ii) contradict each other? No, because the

[1] Lanczos, page 229.

The Lazy Universe. Jennifer Coopersmith, Oxford University Press (2017).
 DOI 10.1093/acprof:oso/9780198743040.001.0001

stars and planets are gravity-wells, and although the matter is confined *spatially* within a well (that is, the spread in q values is small), the *speeds*, corresponding to the p values, increase as matter falls into the well (and so the spread in p values is large). Furthermore, from the Second Law of Thermodynamics, we know that the increase in the p s - a sort of 'heat' - more than compensates for the confinement of the q s. All told, it's clear that *both* p and q are essential variables.

The mathematics input. There are two abstract mathematical 'spaces' that occur in the variational mechanics: there is the configuration space of Lagrangian Mechanics - we could call it 'Narnia'; and the phase space of Hamiltonian Mechanics - we could call it 'Never Never Land'. The surprising thing is that from these abstract spaces (Narnia and Never Never Land) we learn some very real results. This is due to the peculiar advantage of an abstract mathematical space - it is specially contrived to have perfect fiducial properties which no real space could ever have.

Philosophical input. Lanczos reminds us of the crucial importance of philosophical questions but always with the understanding that "when the physics is right then the philosophy is right, no matter how long it takes the philosophers to come around."[2] The true laws of physics must free themselves from the artifice of a particular point of view, and must build up a picture from purely local information. In the final analysis, 'far away' and 'long ago' are mere conjecture. (Descartes anticipated this: despite having discovered coordinates along infinite axes, he believed that only *local* motion made any sense.) The Principle of Least Action satisfies these requirements as it yields an invariant result (the action is least independently of what coordinates are used, what frame of reference is used, and even whether or not the frame of reference is moving); and, although the initial and final states are a finite 'distance' apart, the 'space' is explored incrementally and *locally*, both in 'time' and 'space'. Newton postulates the prior existence of an absolute Space and an absolute Time, whereas Einstein, and the Variational Mechanics, make no such assumptions.

Also related to philosophy, during the eighteenth and nineteenth centuries the Principle fell out of favour because of the teleological question - how does the particle know which path to choose? Although we should not judge one century from another, we can now see that this was a non-question: the particle does not 'choose' a path

[2] *Coopersmith J, EtSC*, 2nd edition, page 356.

anymore than a Newtonian particle 'knows how to calculate the acceleration' in order to 'know where to move to next'. This is confirmed in Feynman's 'many histories' version of quantum mechanics where it is clear that choice doesn't come into it - *all* the possible paths are taken.

The main objection to using the Principle of Least Action is its counter-intuitiveness: it has to do with complicated scalar energy functions rather than simple, vectorial, forces. For sure, in answering certain simple vectorial questions (such as, what tensions are required for equilibrium when three taut cords meet at a point), it is correct to use the Newtonian approach. However, in modern physics the force-analysis has been superseded by an energy-analysis. Forces are crucial to the understanding of simple problems, yes, but just as children use counting numbers while mathematicians have graduated to the use of real numbers, so we must graduate from forces to kinetic energy and to the energy 'structure' functions. Also, in the Variational Mechanics we look for solutions between two pre-determined end-states. This is less intuitive than the forwards-marching directive of Newton's Second Law of Motion. Apparently, Feynman was, at first exposure to the Principle of Least Action, repelled by it.[3] However, our intuition has not always been this way. To Galileo, the idea of a body moving along an unbounded path, heading out into the infinite void, was anathema - all 'natural' motion was in a circle. More fundamentally, we appreciate that there is always a *duality* inherent in the acquisition of knowledge:[4] it is built up, ultimately, by comparing one thing with another - like the definitions of words in a dictionary. We remember that while we paddle and luxuriate in the warm waters of 'common sense', these waters are shallow, and we must be prepared to follow the logic of our mathematical arguments, and train up our intuition accordingly. Hamilton jolts us out of our complacency. For example, when explaining how gravity, so familiar, was yet not at all obvious, he wrote: "Do you think that

[3] James Gleick, "*Genius: The Life and Science of Richard Feynman*", Pantheon Books 1992.

[4] Author's speculation: The 'absolute geometry' (aka differential geometry) assumes this duality. For example, tensors, however many dimensions they may have, they only ever have two kinds of indices (the so-called raising and lowering operators, or contravariant and covariant forms). But perhaps a more general triality, quadrality, ..., may yet be discovered?

we [actually] *see* the [gravitational] attraction of the planets? We scarcely [even] see their orbits."[5]

A truly fundamental theory will have repercussions in all sorts of ways, and in all sorts of previously disconnected areas of enquiry. Hamilton was motivated to explain everything purely algebraically but ended up bringing in a theory that was geometrical and visual. Another example comes from Einstein. His Theory of Special Relativity, designed to cope with speeds near c, resulted in a radically new understanding of energy (the famous $E = mc^2$) that applies at low or even zero speed. Likewise, from the Variational Mechanics we have learned to understand energy in a new more fundamental way, and to realize that 'kinetic energy' and 'potential energy' are, respectively, examples of '*component*-energy' and '*whole-system* energy', but also to appreciate that these are not hard and fast categories, they depend on the frame of reference.

Why does the Principle of Least Action work? It works because of the importance of modelling the system in the right way. Forces and particles are elemental but they are not necessarily (in fact, not usually) the right elements. Instead we need new system-specific elements like, for example, the stress energy tensor. An analogy comes with economics. Many economics theories base themselves just on the needs of individuals. Some even go so far as to deny the existence of society. This is wrong: a society is more than the sum of its individuals. Likewise, a physical system is more than the sum of its parts and moreover those parts are not even necessarily Newtonian.

We have found that the Principle of Least Action applies across all scales, from the realm of the microscopic (actually, tinier than that) to the everyday (classical mechanics, engineering, optics, the transmission of radiation, physical chemistry, statistical mechanics, continuum mechanics), and on to the whole cosmos (gravitation due to stars, planets, black holes, and gravity waves). From Hamilton's optical analogy comes the wisdom, not only about the common ancestry of the classical and quantum worlds, but actually that quantum mechanics is the more fundamental theory. (So, instead of always trying to explain quantum weirdness, we should perhaps be trying to explain classical weirdness.) Dissipation must also ultimately come under the ambit of

[5] W R Hamilton, in Hankins, T L, "*Sir William Rowan Hamilton*", John Hopkins Uinversity Press 1980, p 177.

the Principle of Least Action - after all, nature has no sharp divisions, and, as Lanczos advises, dissipative processes do come in functional form when explained at the microscopic level. Even more apropos, the Second Law of Thermodynamics involves tiny incremental processes and an extremal (the maximizing of entropy). Perhaps the true theory combining action and thermodynamics (the Second Law) still awaits discovery - maybe a reader of this book will pick up the baton?

D'Alembert, at the start of this book, proposed that the universe was a unique truth - provided only that the viewpoint was all-embracing, that is, objective. Einstein refined the idea, and put it forward humbly as a question:

> "What I'm really interested in is whether God could have made the world in a different way; that is, whether the necessity of logical simplicity leaves any freedom at all."[6]

We pose the idea in yet a third way: would the laws of physics end up the same if the historical progression had been different? For example, instead of starting with forces and particles, could we have started directly with energy and action? The Principle of Least Action shows us that the answer is - yes![7] Note that this still doesn't answer Einstein's question; all we can say is that the Principle of Least Action is our best candidate for a necessary theory.

In whatever domain, and on whatever scale, it seems that nature seeks out the flat (it abhors a gradient), and the universe really is as lazy as possible.

[6] Einstein's remark to his assistant, Ernst Straus, see Holton G in "*The Nature of Scientific Discovery*", ed Gingerich, Smithsonian Institution Press, Washington (1975).

[7] For example, we have Lanczos's wisdom that it is kinetic energy that is the true most primitive measure of inertia (Section 3.4); and we find in certain cases that the kinetic energy serves directly as the Riemannian 'distance function' (footnote in Section 7.5, Optional reading in Section 3.6, and the end of Appendix A7.9).

APPENDIX A1.1

Newton's Laws of Motion

Law I

EVERY BODY PERSEVERES IN ITS STATE OF REST, OR OF UNIFORM MOTION IN A RIGHT LINE, UNLESS IT IS COMPELLED TO CHANGE THAT STATE BY FORCES IMPRESSED THEREON.

Law II

THE ALTERATION OF MOTION IS EVER PROPORTIONAL TO THE MOTIVE FORCE IMPRESSED; AND IS MADE IN THE DIRECTION OF THE RIGHT LINE IN WHICH THAT FORCE IS IMPRESSED.

Law III

TO EVERY ACTION THERE IS ALWAYS OPPOSED AN EQUAL REACTION: OR THE MUTUAL ACTION OF TWO BODIES UPON EACH OTHER ARE ALWAYS EQUAL, AND DIRECTED TO CONTRARY PARTS.

Philosophiae naturalis principia mathematica (The Mathematical Principles of Natural Philosophy), by Isaac Newton, familiarly known as *The Principia*, first published in 1687, translated in 1729 by Motte. The Laws occur in Book I.

APPENDIX A2.1

Portraits of the physicists

Pierre de Fermat (1601 or 7–65)
Johann Bernoulli (1667–1748)
Pierre Louis de Maupertuis (1698–1759)
Leonhard Euler (1707–83)
Jean le Rond d'Alembert (1717–83)
Joseph-Louis Lagrange (1736–1813)
William Rowan Hamilton (1805–65)
Carl Gustav Jacobi (1804–51)
Emmy Noether (1882–1935)
Erwin Schrödinger (1887–1961)
Cornelius Lanczos (1893–1974)
Richard Feynman (1918–88)

Permissions

Thank you to the following for permission to reproduce the images:

Pierre de Fermat - ©Jean Lepage, Musée de Narbonne, Ville de Narbonne
William Rowan Hamilton - By permission of the Royal Irish Academy © RIA
Emmy Noether - courtesy of Bryn Mawr College, copyright-holder unknown
Cornelius Lanczos - The Dublin Institute for Advanced Studies
Erwin Schrödinger - Special Collections, Oregon State University, copyright-holder unknown
Richard Feynman - courtesy of the Estate of Richard P Feynman

Remaining images - thank you to Wikimedia Commons, a donation has been made.

Pierre de Fermat
(1601 or 7–65)

Johann Bernoulli
(1667–1748)

Pierre Louis de Maupertuis
(1698–1759)

Leonhard Euler
(1707–83)

Jean le Rond d'Alembert
(1717–83)

Joseph-Louis Lagrange
(1736–1813)

William Rowan Hamilton
(1805–65)

Carl Gustav Jacobi
(1804–51)

Emmy Noether
(1882–1935)

Erwin Schrödinger
(1887–1961)

Cornelius Lanczos
(1893–1974)

Richard Feynman
(1918–88)

APPENDIX A3.1

Reversible displacements

We require that the virtual displacements are reversible, in other words, both $+\delta x$ and $-\delta x$ must be possible. The reason for this requirement is as follows. We are trying to establish a stationary condition and this is usually also an extremum - a maximum or a minimum. However the reverse is not always true - we can have an extremum without it being a stationary point. For example, consider looking at a washing-line through a kitchen window. As seen from inside the kitchen, the curve of the washing-line has a maximum at position *a*, and a minimum at position *b*, but neither of these is stationary (in fact, the stationary position is a minimum and is at *c*, outside the frame).

The conclusion: to rule out false alarms (extrema which don't correspond to stationary points) we must veto points on a boundary, and only investigate points where the displacements can occur in both positive and negative directions.

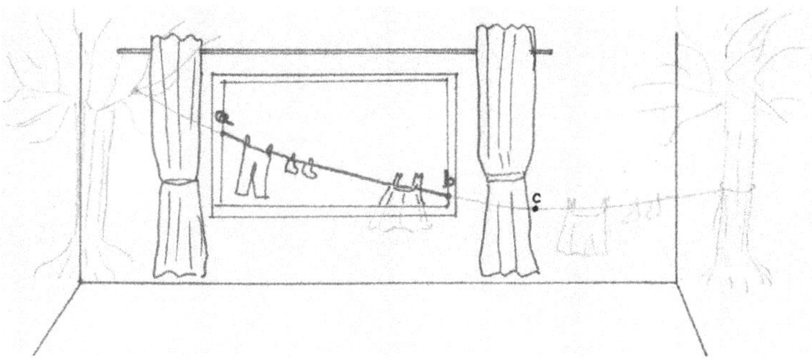

Figure A3.1 Washing-line seen through a window.

APPENDIX A6.1

Worked examples in Lagrangian Mechanics

1) One particle, mass m, potential V, Cartesian, 2-D

Coordinates: $q_1 = x$, $q_2 = y$

$T = \frac{1}{2}m(\dot{x}^2 + \dot{y}^2)$, $\quad V = V(x, y) \qquad$ (note, V is not a function of $\dot{x}$ or $\dot{y}$)

$$L = T - V = \frac{1}{2}m(\dot{x}^2 + \dot{y}^2) - V(x, y)$$

The Lagrange Equations are:

$$\frac{d}{dt}\left(\frac{\partial L}{\partial \dot{q}_i}\right) - \frac{\partial L}{\partial q_i} = 0 \qquad i = 1 \text{ and } 2$$

For $i = 1$, we have: $q_1 = x$, $\partial L/\partial q_1 = -\partial V/\partial x$, $\partial L/\partial \dot{q}_1 = m\dot{x}$ and $\frac{d}{dt}(\partial L/\partial \dot{q}_1) = m\ddot{x}$

Therefore the Lagrange Equation for $i = 1$ is:

$$m\ddot{x} - (-\partial V/\partial x) = 0$$

Rearranging, and noting that $F_x = -\partial V/\partial x$, we obtain:

$$m\ddot{x} = F_x$$

For $i = 2$, in an identical manner, we find:

$$m\ddot{y} = F_y$$

In other words, we have arrived at Newton's Second Law of Motion.

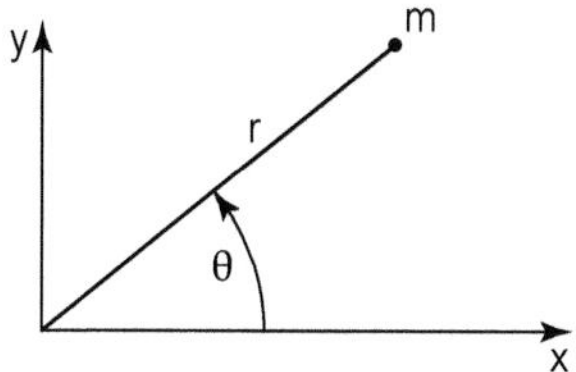

Figure A6.1.1 Plane polar coordinates.

2) One particle, mass m, potential V, plane polar coordinates

Coordinates: $q_1 = r, q_2 = \theta$

transformation equations:

$$x = r\cos\theta$$
$$y = r\sin\theta$$

$$\dot{x} = \dot{r}\cos\theta - r\dot{\theta}\sin\theta$$
$$\dot{y} = \dot{r}\sin\theta + r\dot{\theta}\cos\theta$$

$T = \frac{1}{2}m(\dot{x}^2 + \dot{y}^2)$ becomes $T = \frac{1}{2}m(\dot{r}^2 + r^2\dot{\theta}^2)$

The potential energy is $V = V(r, \theta)$

$L = T - V$ and so $L = \frac{1}{2}m(\dot{r}^2 + r^2\dot{\theta}^2) - V(r, \theta)$

The Lagrange Equations are:

$$\frac{d}{dt}\left(\frac{\partial L}{\partial \dot{q}_i}\right) - \frac{\partial L}{\partial q_i} = 0 \qquad i = 1 \text{ and } 2$$

For $i = 1$:

$$q_1 = r,\ \partial L/\partial r = mr\dot{\theta}^2 - \partial V/\partial r,\ \partial L/\partial \dot{r} = m\dot{r} \text{ and } \tfrac{d}{dt}\left(\partial L/\partial \dot{r}\right) = m\ddot{r}$$

From the Lagrange Equation, $m\ddot{r} - (mr\dot{\theta}^2 - \partial V/\partial r) = 0$

$$\Rightarrow m\ddot{r} - mr\dot{\theta}^2 = -\partial V/\partial r = Q_r$$

For $i = 2$:

$$q_2 = \theta,\ \partial L/\partial\theta = -\partial V/\partial\theta,\ \partial L/\partial\dot{\theta} = mr^2\dot{\theta} \text{ and } \frac{d}{dt}\left(\partial L/\partial\dot{\theta}\right) = mr^2\ddot{\theta} + 2mr\dot{r}\dot{\theta}$$

Substituting into the second Lagrange Equation:

$$mr^2\ddot{\theta} + 2mr\dot{r}\dot{\theta} - (-\partial V/\partial\theta) = 0$$
$$\Rightarrow mr^2\ddot{\theta} + 2mr\dot{r}\dot{\theta} = -\partial V/\partial\theta = Q_\theta$$

Q_r and Q_θ are the generalized forces (Appendix A6.5):

$$Q_r = \mathbf{F}_r\cdot\partial\mathbf{r}_r/\partial r + \mathbf{F}_\theta\cdot\partial\mathbf{r}_\theta/\partial r = \mathbf{F}_r\cdot\partial\mathbf{r}_r/\partial r = F_r \text{ and}$$
$$Q_\theta = \mathbf{F}_r\cdot\partial\mathbf{r}_r/\partial\theta + \mathbf{F}_\theta\cdot\partial\mathbf{r}_\theta/\partial\theta = \mathbf{F}_\theta\cdot r\,\mathbf{n} = rF_\theta$$

(as $\partial\mathbf{r}_\theta/\partial r$ and $\partial\mathbf{r}_r/\partial\theta$ are zero, and $\mathbf{n}$ is a unit vector in the direction of changes in θ).

Collecting all these results together, the equations of motion are:

$$m\ddot{r} - \underbrace{mr\dot{\theta}^2}_{centripetal} = Q_r = F_r \text{ and}$$

$$mr^2\ddot{\theta} + 2mr\dot{r}\dot{\theta} = \frac{d}{dt}(mr^2\dot{\theta}) = Q_\theta = rF_\theta$$

where $mr\dot{\theta}^2$ is the centripetal force, and $mr^2\dot{\theta}$ is the angular momentum - so we have arrived at the standard torque equation of classical mechanics.

Note that examples 1) and 2) are the same scenario but in different coordinates - *but angular momentum and the centripetal force occur only in* 2). Now the coordinate, θ does not occur in T (only $\dot{\theta}$ occurs), and so if V also doesn't depend on θ - that is, we have a central potential, $V = V(r)$ only - then we have L with θ as an 'absent' or 'ignorable' coordinate. We will then find that the angular momentum is a constant of the motion, and so $Q_\theta = 0$.

3) N free particles, mass m_i

In Cartesian coordinates we have:

$$T = \frac{1}{2}\sum_{i=1}^{N} m_i(\dot{x}_i^2 + \dot{y}_i^2 + \dot{z}_i^2) \quad \text{and } V = 0$$

In generalized coordinates we have:

$$T = \frac{1}{2}\sum_{i=1}^{3N} m_i(\dot{q}_i^2) \quad \text{and } V = 0$$

where $m_1 = m_2 = m_3, m_4 = m_5 = m_6$, and so on.

We have $L = T - V = \frac{1}{2}\sum_{i=1}^{3N} m_i(\dot{q}_i^2)$ and the Lagrange Equations are:

$$\frac{d}{dt}\left(\frac{\partial L}{\partial \dot{q}_i}\right) - \frac{\partial L}{\partial q_i} = 0 \qquad i = 1 \text{ to } 3N$$

Now $\partial L/\partial q_i = 0$, $\partial L/\partial \dot{q}_i = m_i\dot{q}_i$, and $\frac{d}{dt}\left(\partial L/\partial \dot{q}_i\right) = m_i\ddot{q}_i$, therefore the Lagrange Equation for any i yields:

$$m_i\ddot{q}_i = Q_i = 0$$

Note how we have gone from N particles in three-dimensional Cartesian space to $3N$ particles in $3N$-dimensional configuration space.

4) Atwood's Machine

Masses m_1 and m_2 are attached at the ends of an inextensible cord which passes without friction over a pulley. The cord and pulley have negligible mass. We can consider two coordinates, x_1 and x_2, and the constraint condition that $x_1 + x_2 = l$, or we can have just one coordinate, x, as shown in the figure.

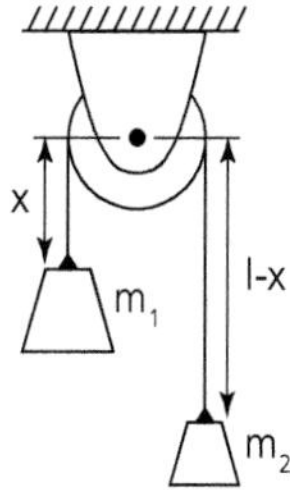

Figure A6.1.2 Atwood's Machine.

We adopt the sign convention that x increases down towards m_1 where $m_1 \geq m_2$.

$$T = \frac{1}{2}(m_1 + m_2)\dot{x}^2, \quad V = -m_1 g x - m_2 g(l - x)$$

(g is the acceleration due to gravity)

$$L = T - V = \frac{1}{2}(m_1 + m_2)\dot{x}^2 + m_1 g x + m_2 g(l - x)$$

Now $\partial L/\partial x = (m_1 - m_2)g$, $\partial L/\partial \dot{x} = (m_1 + m_2)\dot{x}$ and $\frac{d}{dt}\partial L/\partial \dot{x} = (m_1 + m_2)\ddot{x}$

The Lagrange Equation is:

$$\frac{d}{dt}\left(\frac{\partial L}{\partial \dot{q}_i}\right) - \frac{\partial L}{\partial q_i} = 0 \qquad \text{where } q = x$$

$$\Rightarrow (m_1 + m_2)\ddot{x} = (m_1 - m_2)g \Rightarrow \ddot{x} = (m_1 - m_2)g/(m_1 + m_2)$$

This is the familiar Newtonian result, but *at no stage have we had to determine the tension in the cord.*

5) 'Half-Atwood' - or Connected Masses Revisited

We consider the *same problem* as given in Chapter 4 but now analysed using Lagrangian Mechanics instead of the Principle of Virtual Work:

There is one degree of freedom, x, and we adopt the sign convention that x increases downward. We have:

$$T = \frac{1}{2}(m_1 + m_2)\dot{x}^2, \quad V = -m_2 g x$$

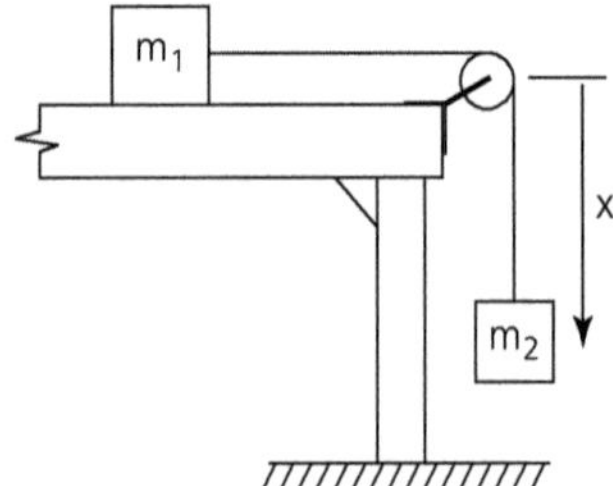

Figure A6.1.3 The 'Half-Atwood' Machine.

(g is the acceleration due to gravity)

$$L = T - V = \frac{1}{2}(m_1 + m_2)\dot{x}^2 + m_2 g x \text{ and therefore}$$

$$\partial L/\partial x = m_2 g,\ \partial L/\partial \dot{x} = (m_1 + m_2)\dot{x},\ \text{and } \frac{d}{dt}\,\partial L/\partial \dot{x} = (m_1 + m_2)\ddot{x}$$

The Lagrange Equation is:

$$\frac{d}{dt}\left(\frac{\partial L}{\partial \dot{q}_i}\right) - \frac{\partial L}{\partial q_i} = 0 \quad \text{where } q = x$$

and therefore $(m_1 + m_2)\ddot{x} - m_2 g = 0$

$$\Rightarrow \ddot{x} = m_2 g/(m_1 + m_2)$$

(Compare with the Newtonian approach:

balance of forces at m_1, $m_1\ddot{\mathbf{x}} = \mathbf{F}_{tension}$
balance of forces at m_2, $m_2\ddot{\mathbf{x}} = m_2\mathbf{g} - \mathbf{F}_{tension}$

eliminate $\mathbf{F}_{tension}$, and then find that $\ddot{x} = m_2 g/(m_1 + m_2)$ as before.)

6) Mass, m, attached to a spring

There is one coordinate, the extension of the spring, x, (it is measured from the unextended length of the spring, x_0). The spring has stiffness constant, k, and we assume that its mass is negligible, that x is 'small' (so Hooke's Law applies), and that there is no dissipation. We have:

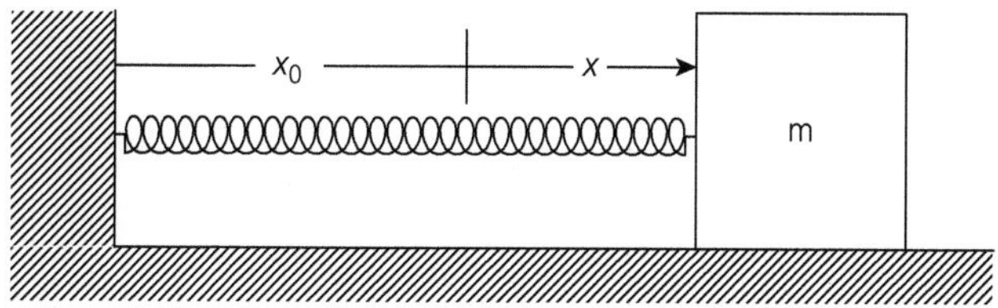

Figure A6.1.4 Mass attached to a spring.

$$T = \frac{1}{2}m\dot{x}^2,\ V = \frac{1}{2}kx^2$$

$$L = T - V = \frac{1}{2}(m\dot{x}^2 - kx^2)$$

$$\partial L/\partial x = -kx,\ \partial L/\partial \dot{x} = m\dot{x},\ \text{and}\ \frac{d}{dt}\partial L/\partial \dot{x} = m\ddot{x}$$

The Lagrange Equation is:

$$\frac{d}{dt}\left(\frac{\partial L}{\partial \dot{q}_i}\right) - \frac{\partial L}{\partial q_i} = 0 \quad \text{where } q = x$$

and so $m\ddot{x} = -kx$ as expected.

7) Planet in orbit around the Sun

Assume the orbit is in a plane, and that the Sun is fixed at the origin of coordinates. We use plane polar coordinates, $q_1 = r$ and $q_2 = \theta$, and have:

$$T = \frac{1}{2}m(\dot{r}^2 + r^2\dot{\theta}^2) \text{ and } V = -GmM/r$$

(m and M are the masses of the planet and the Sun respectively, G is the gravitational constant).

$$L = T - V \text{ and so } L = \frac{1}{2}m(\dot{r}^2 + r^2\dot{\theta}^2) + GmM/r$$

The Lagrange Equations are:

$$\frac{d}{dt}\left(\frac{\partial L}{\partial \dot{q}_i}\right) - \frac{\partial L}{\partial q_i} = 0 \quad i = 1 \text{ and } 2$$

For $i = 1$:

$$q_1 = r,\ \partial L/\partial r = mr\dot{\theta}^2 - GmM/r^2,\ \partial L/\partial \dot{r} = m\dot{r},\ \text{and}\ \frac{d}{dt}\left(\partial L/\partial \dot{r}\right) = m\ddot{r}$$

Therefore, from the first Lagrange Equation,

$$m\ddot{r} = mr\dot{\theta}^2 - GmM/r^2$$

where $mr\dot{\theta}^2$ arises from a centripetal force, and $-GmM/r^2$ arises from $\partial V/\partial r$.

For $i = 2$:

$$q_2 = \theta,\ \partial L/\partial\theta = 0,\ \partial L/\partial\dot{\theta} = mr^2\dot{\theta} \text{ and } \frac{d}{dt}\left(\partial L/\partial\dot{\theta}\right) = mr^2\ddot{\theta} + 2mr\dot{r}\dot{\theta}$$

Therefore, from the second Lagrange Equation:

$$\frac{d}{dt}\left(\partial L/\partial\dot{\theta}\right) = 0 \quad \text{and this implies that } \partial L/\partial\dot{\theta} = \textit{constant}$$

Bearing in mind that $\partial L/\partial\dot{\theta} = mr^2\dot{\theta}$ = angular momentum, then from the second Lagrange Equation we learn that the angular momentum of the planet remains *constant* during its motion - and this is not surprising as the expression for L does not contain θ, in other words, the system is symmetric with respect to rotations, or, equivalently, the system-space is isotropic.

If we are on the planet but are not aware of the change in angle, $\dot{\theta}$ (say, we're in thick fog all the time) then the $\frac{1}{2}mr^2\dot{\theta}^2$ term in L can be considered as a contribution to potential energy, V, rather than to kinetic energy, T. If we're not aware of our orbital motion at all (say we're pre-Copernicans!) then L will still have the same form but all of it will be considered as potential energy. (The strength of gravity will change slightly throughout the planet-year, and a very sensitive Foucault's pendulum could still detect the orbital rotation - note that we are ignoring spin.)

8) Spherical pendulum

$$q_1 = \theta \text{ and } q_2 = \phi$$
$$T = \frac{1}{2}ml^2(\dot{\theta}^2 + (\sin\theta)^2\dot{\phi}^2)$$
$$V = mgl\,(1 - \cos\theta)$$

$$L = T - V = \frac{1}{2}ml^2(\dot{\theta}^2 + (\sin\theta)^2\dot{\phi}^2) - mgl(1 - \cos\theta)$$

The Lagrange Equations are:

$$\frac{d}{dt}\left(\frac{\partial L}{\partial\dot{q}_i}\right) - \frac{\partial L}{\partial q_i} = 0 \quad i = 1 \text{ and } 2$$

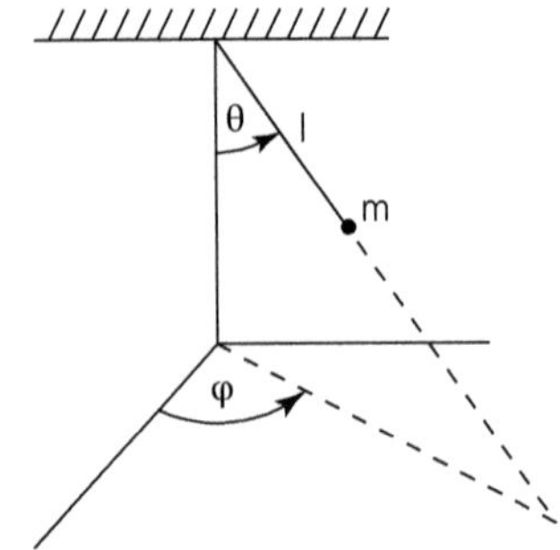

Figure A6.1.5 Spherical pendulum.

For $i = 1$:

$$q_1 = \theta,\ \partial L/\partial\theta = ml^2 \sin\theta \cos\theta\dot{\phi}^2 - mgl\sin\theta$$
$$\partial L/\partial\dot{\theta} = ml^2\dot{\theta},\ \frac{d}{dt}\left(\partial L/\partial\dot{\theta}\right) = ml^2\ddot{\theta}$$

From the Lagrange Equation:

$$ml^2\ddot{\theta} - ml^2 \sin\theta \cos\theta\dot{\phi}^2 + mgl\sin\theta = 0$$
$$\Rightarrow \ddot{\theta} - \sin\theta \cos\theta\dot{\phi}^2 + g\sin\theta/l = 0$$

For $i = 2$:

$$q_2 = \phi,\ \partial L/\partial\phi = 0,\ \partial L/\partial\dot{\phi} = ml^2(\sin\theta)^2\dot{\phi} \text{ and}$$
$$\frac{d}{dt}\left(\partial L/\partial\dot{\phi}\right) = ml^2(2\sin\theta\cos\theta\,\dot{\theta}\dot{\phi} + (\sin\theta)^2\ddot{\phi}) \Rightarrow \ddot{\phi} + 2\dot{\theta}\dot{\phi}/\tan\theta = 0$$

9) Projectile motion, Cartesian coordinates, 2-D (ignoring air resistance)

$$q_1 = x,\ q_2 = y,\ T = \frac{1}{2}m(\dot{x}^2 + \dot{y}^2),\ V = mgy, \text{ therefore:}$$
$$L = T - V = \frac{1}{2}m(\dot{x}^2 + \dot{y}^2) - mgy$$

The Lagrange Equations are:

$$\frac{d}{dt}\left(\frac{\partial L}{\partial\dot{q}_i}\right) - \frac{\partial L}{\partial q_i} = 0 \quad i = 1 \text{ and } 2$$

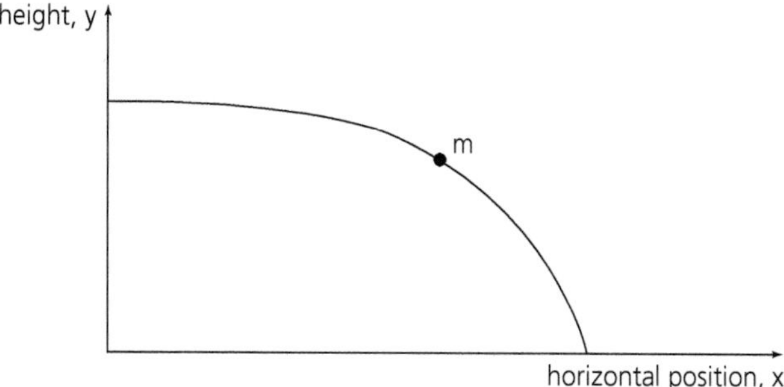

Figure A6.1.6 Projectile motion.

For $i = 1$:

$$\partial L/\partial x = 0,\ \partial L/\partial \dot{x} = m\dot{x},\ \frac{d}{dt}(\partial L/\partial \dot{x}) = m\ddot{x}$$

The Lagrange Equation leads to: $\ddot{x} = 0$

For $i = 2$:

$$\partial L/\partial y = -mg,\ \partial L/\partial \dot{y} = m\dot{y},\ \frac{d}{dt}(\partial L/\partial \dot{y}) = m\ddot{y}$$

The Lagrange Equation leads to: $\ddot{y} = -g$

Solving for x and y we obtain the usual equations:

$$x = v_{x0}t \text{ and } y = v_{y0}t - \frac{1}{2}gt^2$$

where v_{x0} and v_{y0} are the starting speeds in the x and y directions respectively.

10) Bead sliding on a uniformly rotating straight wire (from Goldstein, *Classical Mechanics*, 2nd Edition, page 29)

As the rate of rotation of the wire, ω, is constant, then the system is symmetric (there is no dependence on θ) and so there is just one degree of freedom, $q = r$, the position of the bead along the wire. The bead has mass m, and the wire is assumed to have a negligible mass. Also, we assume that the bead slides without friction.

We have:

$$T = \frac{1}{2}m\dot{r}^2 + \frac{1}{2}mr^2\omega^2$$

(Note that the second term in T has *no* dependency on speed, $\dot{r}$, squared or otherwise.)

We have the constraint that the bead must stay on the wire but there is no external force-field (that is, $V = 0$). So we have:

$$L = T - V = T = \frac{1}{2}m\dot{r}^2 + \frac{1}{2}mr^2\omega^2 \text{ and}$$

$$\partial L/\partial r = mr\omega^2,\ \partial L/\partial \dot{r} = m\dot{r},\ \text{and } \frac{d}{dt}(\partial L/\partial \dot{r}) = m\ddot{r}$$

The Lagrange Equation is:

$$\frac{d}{dt}\left(\frac{\partial L}{\partial \dot{q}}\right) - \frac{\partial L}{\partial q} = 0 \quad q = r$$

$$\Rightarrow m\ddot{r} - mr\omega^2 = 0 \Rightarrow \ddot{r} = r\omega^2$$

In other words, the bead accelerates outwards along the wire (in the direction of increasing r), and this acceleration depends on both r and ω^2.

Note that *at no stage have we needed to know the force that constrains the bead to stay on the wire*. Also, if we change to a different reference frame sited on the bead, then $\dot{r} = 0$ and so the kinetic energy is zero, but the bead will 'feel' as if there's a force acting on it which steadily increases with time. In effect, we go from $V = 0$ to $T = 0$ depending on the reference frame.

11) Particle with mass m, charge e, in an electromagnetic field (Cartesian coordinates)

The coordinates are x, y, and z making up the velocity vector, $\mathbf{v}$. The Lagrangian function is:

$$L = \frac{1}{2}mv^2 - e\phi + e\mathbf{A} \cdot \mathbf{v}$$

where ϕ is the scalar potential associated with the electric field, $\mathbf{E}$, and $\mathbf{A}$ is the vector potential associated with the magnetic field, $\mathbf{B}$. The Lagrange Equations are:

$$\frac{d}{dt}\left(\frac{\partial L}{\partial \dot{q}_i}\right) - \frac{\partial L}{\partial q_i} = 0 \quad q_1 = x,\ q_2 = y,\ q_3 = z$$

This problem is continued in an analysis using Hamilton's Mechanics, see Appendix A7.3. Note the important advance: although *the potential does depend on velocity* (on both speed and direction of motion), a suitable Lagrangian can be found, and the Principle of Least Action can be applied in the usual way.

12) Electrical and mechanical systems compared

We assume the idealizations implicit in a circuit diagram, and also that the inductances are independent of the current going through them (e.g. they don't have iron cores). The charges $Q_1, Q_2, \ldots$ and currents $I_1, I_2, \ldots$ are, respectively, the generalized coordinates, $q_1, q_2, \ldots$ and generalized speeds, $\dot{q}_1, \dot{q}_2, \ldots$. Now, because the total charge must be conserved, then not all the currents can be independently chosen, and it is Kirchoff's Laws which provide the equations of constraint. We have:

$$L = T - V \text{ where}$$

$$T = \frac{1}{2}\sum_i L_i \dot{q}_i^2 + \frac{1}{2}\sum_{ik, i\neq k} M_{ik}\dot{q}_i\dot{q}_k \text{ and}$$

$$V = \sum_i q_i^2/(2C_i) - \sum_i \varepsilon_i(t) q_i$$

where the L_i are self-inductances, M_{ik} are mutual inductances, C_i are capacitances, and $\varepsilon_i(t)$ are time-dependent external emfs. Also, we have dissipative forces, F_i^{diss},

$$F_i^{diss} = -\partial \mathcal{F}/\partial \dot{q}_i$$

where $\mathcal{F}$ is the (speed-dependent) dissipation function, $\frac{1}{2}\sum_i R_i\dot{q}_i^2$, and the R_i are resistances.

The Lagrange Equations are:

$$\frac{d}{dt}\left(\frac{\partial L}{\partial \dot{q}_i}\right) - \frac{\partial L}{\partial q_i} = F_i^{diss}$$

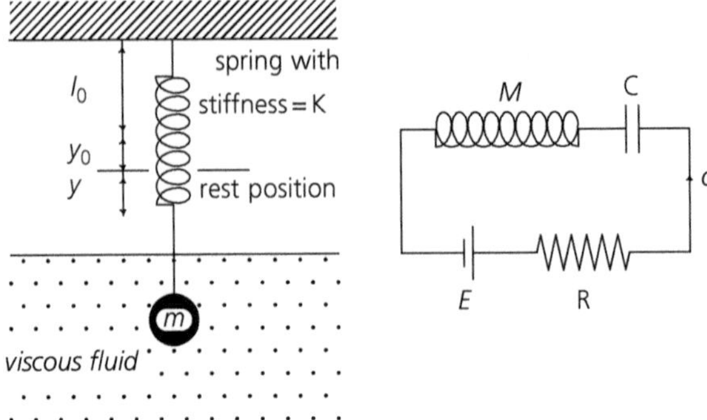

Figure A6.1.7 Electrical and mechanical systems compared.

Note how the dissipative force introduces a 'right-hand side' to the Lagrange Equations (see Appendix A6.5).

As these equations are universally applicable, we can apply the same analysis, and gain the same insights, for an electrical or a mechanical system - *including* dissipation (adapted from Dare Wells, *Lagrangian Dynamics*, Schaum Outline Series, 1967, p309):

Mechanical	Electrical
$L = \frac{1}{2}m\dot{y}^2 - \frac{1}{2}k(y + y_0)^2 + mgy$	$L = \frac{1}{2}M\dot{q}^2 - \frac{1}{2}q^2/C + \varepsilon q$
Dissipative force:	
viscous drag $= -\alpha\dot{y}$	electrical resistive force $= -R\dot{q}$
the equations of motion are:	
$m\ddot{y} + ky = -\alpha\dot{y}$	$M\ddot{q} + q/C = -R\dot{q}$
(as $mg = ky_0$)	

Note that Lagrangian Mechanics is here able to cope with dissipative effects, as these are modelled in a *functional* form. Note also that components in an electrical circuit can now be considered as having massy inertial attributes.

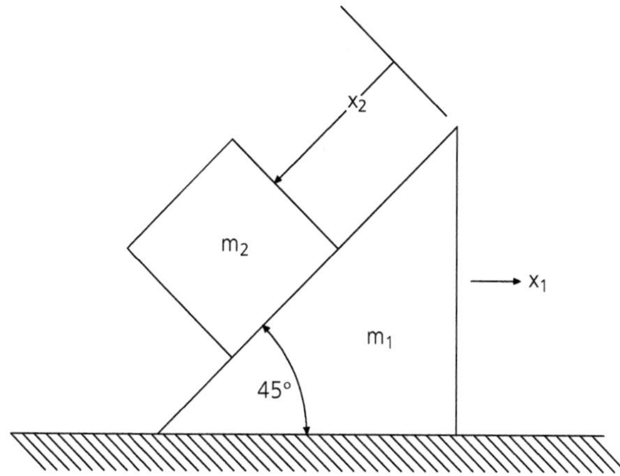

Figure A6.1.8 One block sliding on another block (adapted from Donald Greenwood, *Classical Dynamics*, Dover, 1977, p59).

13) One block sliding on another block, thereby setting the latter in motion

A block of mass, m_2, slides, without friction, down the inclined surface of another block, m_1, which in turn slides without friction along a horizontal surface. The generalized coordinates are $q_1 = x_1$ and $q_2 = x_2$. We have:

$$L = T - V = \frac{1}{2}m_1\dot{x}_1^2 + \frac{1}{2}m_2(\dot{x}_1^2 + \dot{x}_2^2 - \sqrt{2}\dot{x}_1\dot{x}_2) - (-m_2 g x_2/\sqrt{2})$$

(g is the acceleration due to gravity)

The Lagrange Equations are:

$$\frac{d}{dt}\left(\frac{\partial L}{\partial \dot{q}_i}\right) - \frac{\partial L}{\partial q_i} = 0 \quad i = 1 \text{ and } 2$$

For $i = 1$: $\partial L/\partial x_1 = 0$, $\partial L/\partial \dot{x}_1 = (m_1 + m_2)\dot{x}_1 - m_2\dot{x}_2/\sqrt{2}$, and

$$\frac{d}{dt}(\partial L/\partial \dot{x}_1) = (m_1 + m_2)\ddot{x}_1 - m_2\ddot{x}_2/\sqrt{2}$$

From the Lagrange Equation, $(m_1 + m_2)\ddot{x}_1 - m_2\ddot{x}_2/\sqrt{2} = 0 \Rightarrow$

$$\ddot{x}_1 = m_2\ddot{x}_2/\sqrt{2}(m_1 + m_2) \tag{1}$$

(Note that x_1 doesn't occur in L and is therefore an 'ignored coordinate', and so the momentum in the x_1 direction remains constant.)

For $i = 2$: $\partial L/\partial x_2 = m_2 g/\sqrt{2}$, $\partial L/\partial \dot{x}_2 = m_2\dot{x}_2 - m_2\dot{x}_1/\sqrt{2}$, and

$$\frac{d}{dt}(\partial L/\partial \dot{x}_2) = m_2(-\ddot{x}_1/\sqrt{2} + \ddot{x}_2)$$

From the Lagrange Equation, $m_2(-\ddot{x}_1/\sqrt{2} + \ddot{x}_2 - m_2 g/\sqrt{2} = 0 \Rightarrow$

$$\sqrt{2}\ddot{x}_2 - \ddot{x}_1 = g \tag{2}$$

Solving (1) and (2) we obtain:

$$\ddot{x}_2 = \sqrt{2}g(m_1 + m_2)/(2m_1 + m_2) \quad \text{and} \quad \ddot{x}_1 = gm_2/(2m_1 + m_2)$$

Optional

What is the force between the blocks? Greenwood determines this using the method of Lagrange Multipliers (Appendix A6.4). He says that the force acts at right angles to the inclined slope, in a new direction, x_3. He finds that this force has magnitude:

$$\text{generalized force} = \lambda = \sqrt{2}gm_1m_2/(2m_1 + m_2)$$

This is a constraint force and is compressive in nature (it stops m_2 from sinking into the surface of m_1). The derivation is curious as it relies upon the constraint condition, $\dot{x}_3 = 0$, and yet $\dot{x}_3$ does nowhere appear in the Lagrangian (equation (2-79) in Greenwood).

APPENDIX A6.2

Proof that T is a function of v^2

The author gratefully acknowledges the priority of Maimon[1].

We hypothesize that for any particle of mass, m, and speed, v, there exists a universal kinetic energy function, T, and that it depends on m and v in some, as yet, unknown way,[2] $T = T(m, v)$.

Consider a ball of putty of mass, m, and speed, v. If the ball is brought to a standstill by crashing into a wall then, by energy conservation, all of its kinetic energy must be converted into heat (we ignore damage to the wall, sound energy, and so on). Therefore we have: $T(m, v) = \Delta heat$. Also, we have the empirical knowledge that the amount of heating is proportional to the mass of the ball (a ball of n times the mass will generate n times the amount of heating). So we must have:

$$T(m, v) = \Delta heat = mT(v) \qquad \text{(A6.2.1)}$$

This result shows that kinetic energy can be calibrated by a method which does *not* depend on motion - for example, we could use a thermometer.

We now consider an inelastic collision between two identical balls of putty, each of mass m, and each approaching the other with speed v, (see Table I). We view this collision from two reference frames, one at rest, Σ^{rest}, and one moving at speed v in the same direction as one of the balls, Σ'. By the Galilean and Einsteinian Principles of Relativity, energy conservation will apply in each reference frame.

By symmetry: the balls must stick together after the collision (and this is then true in any reference frame); and the total heating must be equal to twice the heating of one ball (in effect each ball acts as a 'wall' for the other). Also, by Einstein's Principle of Relativity, the same physics must occur in all reference frames, and so we must have $\Delta heat' = \Delta heat$ (we are also assuming that the mass m, is independent of the reference frame).

Conservation of energy in the moving reference frame requires that

$$\begin{array}{ccc} \text{before collision} & & \text{after collision} \\ mT(2v) & = & 2mT(v) + \Delta heat' \end{array} \qquad \text{(A6.2.2)}$$

[1] Ron Maimon, physics.stackexchange.com/questions/535/

[2] The notation $F = F(a, b, c, \ldots)$ means that F is some function of $a, b, c, \ldots$.

Table A6.2 I Maimon: Inelastic collision viewed from two reference frames.

Reference Frame	Before Collision	After Collision
Σ^{rest}	$E_{before} = mT(v) + mT(v) = 2mT(v)$	$E_{after} = \Delta$ heat
Σ' → v	$E'_{before} = mT(2v)$	$E'_{after} = 2mT(v) + \Delta$ heat′

But we also know that:

$$2mT(v) + \Delta heat' = 2mT(v) + \Delta heat = 2mT(v) + 2mT(v) \qquad \text{(A6.2.3)}$$

Therefore, combining equations (A6.2.2) and (A6.2.3), we have:

$$mT(2v) = 2mT(v) + 2mT(v) \qquad \text{(A6.2.4)}$$

or

$$T(2v) = 4T(v) \qquad \text{(A6.2.5)}$$

The only way this can be true is if T is a *quadratic* function of v, in other words, we *must* have $T = T(v^2)$. Finally, bringing back the dependence on m, we must have:

$$T = mT(v^2) \qquad \text{(A6.2.6)}$$

This is fine as far as balls of putty go, but we started out wanting to determine the kinetic energy function for *particles*. We can imagine conducting the collision again and again, each time using balls of putty that are progressively smaller and smaller - but they can never become particles as particles can't heat up, and can't act as 'walls'. To remedy this, we consider another proof, due to Ehlers, Rindler, and Penrose.[3] This proof involves only particles,

[3] Energy conservation as the basis of relativistic mechanics, II, J Ehlers, W Rindler and R Penrose, American Journal of Physics **33** (1965) 995.

Table A6.2 II Ehlers et al. Two systems (the second rotated thro' 90°) viewed from two frames (the second moving at speed v_x).

Frame	schematic	particle	$(\text{speed})^2$	Total energy
Σ^{rest}	→←	1 2	$v_1^2 = v^2$ $v_2^2 = v^2$	$mT(v_1^2) + mT(v_2^2)$
	↓ ↑	3 4	$v_3^2 = v^2$ $v_4^2 = v^2$	$mT(v_3^2) + mT(v_4^2)$
$\Sigma'_{\rightarrow} v_x$	→←	1 2	$v_1^2 = (v - v_x)^2$ $v_2^2 = (v + v_x)^2$	$mT(v_1^2) + mT(v_2^2)$
	↙ ↖	3 4	$v_3^2 = v^2 + v_x^2$ $v_4^2 = v^2 + v_x^2$	$mT(v_3^2) + mT(v_4^2)$ $= 2mT(v^2 + v_x^2)$

and only kinetic energy (although the authors in principle allow for internal changes-of-state, in fact their proof doesn't involve these).

Ehlers et al adopt a different tack to Maimon: instead of comparing *one* system before and after a collision, they compare *two* systems each before the impending collision. They then require that the total before-collision energy is identical, whatever the reference frame. (Note, we shall be following their non-relativistic proof.) They consider, first, a system in which two identical free particles (called 1 and 2), each of mass, m, approach each other with equal speeds, v, along the x-axis; and then another system in which the same particles (now called 3 and 4) approach each other with equal speeds, v, along the y-axis (Table II). They look at these systems from a stationary reference frame, Σ^{rest}, and from a reference frame, Σ', moving at uniform speed, v_x, in the positive x-direction. The speed-components for particles 1 and 2, and for particles 3 and 4, are given in Table II.

All the speed-components have been given as squares, but there is no need to take the square root - we simply define the kinetic energy as some function of speed-squared and see if this leads to a contradiction.

Now, viewed from Σ', the total energy in system (1,2) is $mT(v_1^2) + mT(v_2^2)$ and the total energy in system (3,4) is $2mT(v^2 + v_x^2)$. We also know that $(v^2 + v_x^2) = \frac{1}{2}(v_1^2 + v_2^2)$ and therefore the total energy in system (3,4) may be written as $2mT(\frac{1}{2}(v_1^2 + v_2^2))$. Finally, we also require that the total energies in the two systems are identical (as they were identical when viewed from Σ^{rest}). Therefore, (cancelling out the masses, for now) we must have:

$$T(v_1^2) + T(v_2^2) = 2T\left(\frac{1}{2}(v_1^2 + v_2^2)\right) \tag{A6.2.7}$$

or

$$\frac{1}{2}\left(T(v_1^2) + T(v_2^2)\right) = T\left(\frac{1}{2}(v_1^2 + v_2^2)\right) \tag{A6.2.8}$$

In words, this equation reads: "the average of T at v_1^2, and T at v_2^2, equals T at the average of v_1^2 and v_2^2". In the usual way (continued halving and the requirement that T is continuous) it follows that T *must* be a *linear function of* v^2:

$$T = \tfrac{1}{2}mv^2 + T_0 \tag{A6.2.9}$$

where the 'slope' and 'intercept' are the constants $\frac{1}{2}m$ and T_0 respectively. The factor of $\frac{1}{2}$ ensures that this result coincides with the relativistic result in the limit of low speeds.

It is very interesting that the proof has shown that the *kinetic* energy includes a term, T_0, which is *not* dependent on speed: in other words, even as the speed becomes zero, there is still some 'kinetic' energy left over. In the non-relativistic case this zero-point of kinetic energy is usually normalized to zero but, in the relativistic case, the authors point out that it is no longer arbitrary but is *determined* by the theory. Readers will feel a shiver of recognition to learn that $T_0 = mc^2$ (where, here, m stands for the so-called rest-mass).

This paper is nevertheless, like Maimon's, not applicable to 100 per cent of cases. There is a caveat[4] in which it is required that the total energy is independent of the speed of the reference frame (and indeed it is evident that v_x does not occur in the equations for T). However, this is wrong: famously, in the case of a single free particle, or many free particles moving in parallel and in the same direction, then the system energy *is* dependent on the speed of the reference frame. In these especially simple cases there is no escape, one must simply *define* the energy of each particle to be $T = \frac{1}{2}mv^2$ (granted that, in the non-relativistic case, T_0 may be normalized to zero).

Optional

This quadratic form for kinetic energy is, moreover, consistent with the requirement that it has dual 'dimensionality', that is, it refers to *two* configuration spaces, one containing the covariant vector, $\mathbf{v} = v_i\mathbf{e}^i$, the other containing the contravariant vector, $\mathbf{v} = v^i\mathbf{e}_i$ (using tensor notation and Einstein's

[4] (after Assumption II in Ehlers et al - see earlier footnote for the reference)

summation convention). Furthermore, in those occasional cases where T has a linear dependence on speed (main text, end of section 6.5.1), the externally imposed kinematic condition is usually also linear in speed, and so *together* the *dual* 'dimensionality' is maintained.

Note also, from W Rindler's book "*Special Relativity*", Oliver and Boyd (1969), that conservation of 4-momentum means that the ratios of all the ms are determined (page 85); and that it is an empirical and not fully understood fact that m is always positive (page 87).

APPENDIX A6.3

Energy conservation and the homogeneity of time

We consider a case where the Lagrangian has no explicit dependence on time. Then we perform a translation of the time coordinate, $t \mapsto \tau$, where $\tau = t - \epsilon$, and where ϵ is some small constant. Invariance of the action principle before and after the translation implies that:

$$\delta \int_{t_a}^{t_b} L(q_i(t), \dot{q}_i(t))\, dt = \delta \int_{\tau_a}^{\tau_b} \bar{L}(q_i(\tau), q'_i(\tau))\, d\tau = 0 \qquad \text{(A6.3.1)}$$

($\dot{q}$ means $\dfrac{d(q)}{dt}$, q' means $\dfrac{d(q)}{d\tau}$, and $\bar{L}$ refers to the transformed Lagrangian).

Next we consider a more general time translation in which ϵ itself is a function of time, $\epsilon = \epsilon(\tau)$. We stipulate that $\epsilon(\tau)$ is infinitesimal, continuous, and that at the boundaries t is not transformed, in other words, $t_a = \tau_a$ and $t_b = \tau_b$, or $\epsilon(\tau_a) = \epsilon(\tau_a) = 0$.

From $t = \tau + \epsilon(\tau)$ we deduce that $\frac{d(t)}{d\tau} = 1 + \frac{d(\epsilon)}{d\tau}$, and therefore that:

$$dt = (1 + \epsilon')d\tau \qquad \text{(A6.3.2)}$$

(remembering that $'$ is a shorthand for $\frac{d}{d\tau}$). Also, we have $\dot{q} = \frac{d(q)}{dt} = \frac{d(q)}{d\tau}\frac{d(\tau)}{dt} = q'(1+\epsilon')^{-1}$. Therefore, to first order, we have:

$$\dot{q} = q'(1 - \epsilon') \qquad \text{(A6.3.3)}$$

Substituting τ for t, and $\frac{d}{d\tau}$ for $\frac{d}{dt}$, we find that $L(q_i, \dot{q}_i)$ becomes $\bar{L}(q_i, q'_i(1-\epsilon'))$. As ϵ is small then ϵ' also is small and $\bar{L}$ may be expanded in a Taylor series expansion in ϵ' which, to first order, gives:

$$\bar{L}(q_i, q'_i(1-\epsilon')) = \bar{L}(q_i, q'_i) - \left(\sum_i \left(\frac{\partial \bar{L}}{\partial q'_i}\right) q'_i\right)\epsilon' \qquad \text{(A6.3.4)}$$

Also, in the action integral, dt becomes $(1+\epsilon')d\tau$. Collecting all these parts together, and ignoring 2nd- and higher-order terms, we finally arrive at:

$$\delta A = \delta \int_{t_a}^{t_b} L(q_i, \dot{q}_i)\, dt$$
$$= \delta \int_{\tau_a}^{\tau_b} \bar{L}(q_i, q'_i)\, d\tau - \delta \int_{\tau_a}^{\tau_b} \left[\left(\sum_i \left(\frac{\partial \bar{L}}{\partial q'_i} \right) q'_i \right) - \bar{L} \right] \epsilon' d\tau = 0 \quad \text{(A6.3.5)}$$

Now, from the Action Principle, we know that $\delta \int_{t_a}^{t_b} L(q_i, \dot{q}_i)\, dt$ and $\delta \int_{\tau_a}^{\tau_b} \bar{L}(q_i, q'_i)\, d\tau$ are already guaranteed to be zero. So, we are left with a requirement just involving the last term:

$$\delta \int_{\tau_a}^{\tau_b} \left[\left(\sum_i \left(\frac{\partial \bar{L}}{\partial q'_i} \right) q'_i \right) - \bar{L} \right] \epsilon' d\tau = 0 \quad \text{(A6.3.6)}$$

We use our old trick[1] - integration by parts - and arrive at:

$$\delta A = \delta \int_{\tau_a}^{\tau_b} \frac{d}{d\tau} \left[\left(\sum_i \left(\frac{\partial \bar{L}}{\partial q'_i} \right) q'_i - \bar{L} \right) \epsilon \right] d\tau$$
$$- \delta \int_{\tau_a}^{\tau_b} \frac{d}{d\tau} \left(\sum_i \left(\frac{\partial \bar{L}}{\partial q'_i} \right) q'_i - \bar{L} \right) \epsilon\, d\tau = 0 \quad \text{(A6.3.7)}$$

The first integral contains nothing apart from a total differential, and so it becomes a boundary term:

$$\left[\left(\sum_i \left(\frac{\partial \bar{L}}{\partial q'_i} \right) q'_i - \bar{L} \right) \epsilon \right]_{\tau_a}^{\tau_b} \quad \text{(A6.3.8)}$$

However, because of our condition, $\epsilon(\tau_a) = \epsilon(\tau_a) = 0$, then this boundary term is zero. Finally, we are left with just the second integral in (A6.3.7):

$$\delta \int_{\tau_a}^{\tau_b} \frac{d}{d\tau} \left(\sum_i \left(\frac{\partial \bar{L}}{\partial q'_i} \right) q'_i - \bar{L} \right) \epsilon\, d\tau = 0 \quad \text{(A6.3.9)}$$

[1] (Chapter 4 and the beginning of Chapter 6)

The integral is preceded by the variation symbol, δ, but what is being varied? It is the infinitesimal function, ϵ, which acts as a 'variation', and because this variation is arbitrary (the function ϵ can have any form - provided it is infinitesimal, continuous, and disappears at the boundaries), then it must be the coefficient of ϵ which vanishes. In other words, in order that the integral is zero for *arbitrary* infinitesimal variations, it is necessary that:

$$\frac{d}{d\tau}\left(\sum_i \left(\frac{\partial \bar{L}}{\partial q_i'}\right) q_i' - \bar{L}\right) = 0 \qquad \text{(A6.3.10)}$$

which means that:

$$\left(\sum_i \left(\frac{\partial \bar{L}}{\partial q_i'}\right) q_i' - \bar{L}\right) = \text{constant} \qquad \text{(A6.3.11)}$$

Without loss of generality, we may assert this same result in t coordinates:

$$\left(\sum_i \left(\frac{\partial L}{\partial \dot{q}_i}\right) \dot{q}_i - L\right) = \text{constant} \qquad \text{(A6.3.12)}$$

This conserved quantity has units of energy. No assumptions have been made about L except that it is independent of the time. However, when we consider the simplest default form for L (T is quadratic in the speed coordinates, V depends only on the position coordinates, and $L = T - V$), then it turns out (see Section 6.7, Chapter 6) that $\left(\sum \left(\frac{\partial L}{\partial \dot{q}_i}\right) \dot{q}_i - L\right) = E$, where E is the total energy.

Thus, assuming the validity of the Principle of Least Action, and assuming the homogeneity of time (the requirement of invariance following a time transformation) has resulted in the conservation of energy. This is true even when the time-transformation is a time-dependent one!

APPENDIX A6.4

The method of Lagrange Multipliers

Suppose we have a system of n generalized coordinates, $q_1, q_2, \ldots q_n$, and a function, $F = F(q_1, q_2, \ldots q_n)$. From Section (3.7), Chapter 3, we know that in order to find the stationary value of a function, F, we set its 'variation', δF, to zero:

$$\delta F = \frac{\partial F}{\partial q_1}\delta q_1 + \frac{\partial F}{\partial q_2}\delta q_2 + \ldots + \frac{\partial F}{\partial q_n}\delta q_n = 0 \tag{A6.4.1}$$

Now it is possible that the coordinates may be subject to some extra ('auxiliary') conditions. For example, in the case of one particle in 3 rectangular coordinates, the particle may be constrained to lie on the surface of a certain sphere of radius, r:

$$x^2 + y^2 + z^2 = r^2 \tag{A6.4.2}$$

The number of truly independent coordinates has thereby been reduced from 3 to 2, and we could therefore use (A6.4.2) to express one of the coordinates in terms of the other two, for example

$$y = \sqrt{r^2 - (x^2 + z^2)} \tag{A6.4.3}$$

This isn't very satisfactory, as, from the symmetry of (A6.4.2), there is no reason to single out one coordinate as the dependent one. An alternative strategy - due to Lagrange, in 1788 - is as follows.

First, we re-express the auxiliary condition in the form of a function, f, set to zero,

$$f = f(q_1, q_2, \ldots q_n) = 0 \tag{A6.4.4}$$

(For example, in the case of the condition in (A6.4.2), we have the function $f = (x^2 + y^2 + z^2) - r^2 = 0$.) Next, we take the variation of the condition equation:

$$\delta f = \frac{\partial f}{\partial q_1}\delta q_1 + \frac{\partial f}{\partial q_2}\delta q_2 + \ldots + \frac{\partial f}{\partial q_n}\delta q_n = 0 \tag{A6.4.5}$$

We now have two expressions of identical form - the stationarity of F (equation (A6.4.1)), and the variation of the condition (equation (A6.4.5)) - may we not combine them? Yes, but first we multiply both sides of (A6.4.5) by a constant, λ,

$$\lambda\, \delta f = \lambda \left(\frac{\partial f}{\partial q_1} \delta q_1 + \frac{\partial f}{\partial q_2} \delta q_2 + \ldots + \frac{\partial f}{\partial q_n} \delta q_n \right) = 0 \qquad \text{(A6.4.6)}$$

and then add this to (A6.4.1):

$$\frac{\partial F}{\partial q_1} \delta q_1 + \ldots + \frac{\partial F}{\partial q_n} \delta q_n + \lambda \left(\frac{\partial f}{\partial q_1} \delta q_1 + \ldots + \frac{\partial f}{\partial q_n} \delta q_n \right) = 0 \qquad \text{(A6.4.7)}$$

We have, at first glance, merely rescaled zero back to zero, and then added this zero to δF. However, this step is not trivial, as we have added a *sum* of terms, and only the *whole sum* is necessarily zero. More compactly, (A6.4.7) may be written as:

$$\sum_{i=1}^{n} \left(\frac{\partial F}{\partial q_i} + \lambda \frac{\partial f}{\partial q_i} \right) \delta q_i = 0 \qquad \text{(A6.4.8)}$$

Only $(n-1)$ of our q_i s are independent; and we could choose to eliminate one of them, say, the nth one, q_n. But, rather than eliminating q_n, we could instead - Lagrange's flash of genius - specially select the value of λ in order to make the bracket multiplying δq_n vanish:

$$\left(\frac{\partial F}{\partial q_n} + \lambda \frac{\partial f}{\partial q_n} \right) = 0 \qquad \text{(A6.4.9)}$$

This dispenses with the need to eliminate q_n, and we now have a sum over only $(n-1)$ terms:

$$\sum_{i=1}^{n-1} \left(\frac{\partial F}{\partial q_i} + \lambda \frac{\partial f}{\partial q_i} \right) \delta q_i = 0 \qquad \text{(A6.4.10)}$$

All these δq_i s are now independent of each other, and may be chosen arbitrarily. We therefore have a free variation problem, in which the coefficient of each δq_i, for $i = 1$ to $n-1$, must vanish:

$$\left(\frac{\partial F}{\partial q_i} + \lambda \frac{\partial f}{\partial q_i} \right) = 0 \quad i = 1 \text{ to } n-1 \qquad \text{(A6.4.11)}$$

But now these conditions, (A6.4.11), combined with our one earlier condition, (A6.4.9), lead us right back where we started - to *one* sum over *all n*, (A6.4.8), and where all the brackets must vanish (as if all the δq_i s were free variations after all).

The net result is that we can recast the whole problem and consider it as one involving the variation of a new function, $\bar{F}$, defined as:

$$\bar{F} = F + \lambda f \tag{A6.4.12}$$

where

$$\delta\bar{F} = \delta(F + \lambda f) = \delta F + \delta\lambda\, f + \lambda\, \delta f \tag{A6.4.13}$$

which reduces to

$$\delta\bar{F} = \delta(F + \lambda f) = \delta F + \lambda\, \delta f \tag{A6.4.14}$$

as the term $\delta\lambda\, f$ vanishes on account of our initial auxiliary condition, $f = 0$. We have thus converted a variation problem with an auxiliary condition into a free variation problem with no auxiliary condition. The price that has been paid is that the freely varied function is now one with an extra parameter (degree of freedom), λ.

The method can be extended to the variation of integrals, in which F is now the integrand, say, the Lagrangian, L, and $\bar{F}$ is the modified Lagrangian, $\bar{L}$. Also, we can have many (say, m) auxiliary conditions, $f_1, f_2, \ldots, f_m$, each scaled by a different λ-parameter:

$$\delta\bar{A} = \delta\int_{t_a}^{t_b} \bar{L}\, dt = \delta\int_{t_a}^{t_b} (L + \lambda_1 f_1 + \lambda_2 f_2 + \ldots + \lambda_m f_m)\, dt \tag{A6.4.15}$$

As we are now concerned with an integral, the λ_js ($j = 1$ to m) as well as the q_is ($i = 1$ to n) may all be functions of time.

In summary, the method of Lagrange multipliers has treated a problem with m auxiliary conditions in the following way: instead of eliminating surplus (non-independent) variables, the Lagrangian is modified by the addition of the conditions, each one scaled by an undetermined multiplier, λ_j. The problem is then treated as a free variation (no auxiliary conditions) of this modified Lagrangian, $\bar{L}$. The λ_js are determined afterwards, as functions of t, by demanding that they satisfy the m auxiliary conditions (which now come into play); likewise the q_is are determined as functions of t by demanding that they satisfy the n Lagrange Equations.

APPENDIX A6.5

Generalized Forces

What is the connection between the potential energy function, V, the force, $\mathbf{F}$, and the generalized force, Q? The virtual work, for N particles, may be transformed from rectangular to generalized coordinates as follows:

$$\sum_{i=1}^{N} \mathbf{F}_i \cdot \delta \mathbf{r}_i = \sum_{j=1}^{M} \left[\sum_{i=1}^{N} \left[\mathbf{F}_i \cdot \frac{\partial \mathbf{r}_i}{\partial q_j} \right] \right] \delta q_j = \sum_{j=1}^{M} Q_j \delta q_j \qquad \text{(A6.5.1)}$$

Therefore the generalized forces, Q_j, are given by:

$$Q_j = \sum_{i=1}^{N} \left[\mathbf{F}_i \cdot \frac{\partial \mathbf{r}_i}{\partial q_j} \right] \qquad j = 1 \text{ to } M \qquad \text{(A6.5.2)}$$

Note how the upper limits, N, or M, are not the same. In the especially simple case of a 'central potential', that is, $V = V(\mathbf{r}_1, \mathbf{r}_2, \ldots \mathbf{r}_N)$, we have:

$$\mathbf{F}_i = -\frac{\partial V}{\partial \mathbf{r}_i} \qquad i = 1 \text{ to } N, \quad \text{and}$$

$$Q_j = \sum_{i=1}^{N} \left[-\frac{\partial V}{\partial \mathbf{r}_i} \cdot \frac{\partial \mathbf{r}_i}{\partial q_j} \right] = -\frac{\partial V}{\partial q_j} \qquad j = 1 \text{ to } M \qquad \text{(A6.5.3)}$$

More generally we have $V = V(\mathbf{r}_1, \mathbf{r}_2, \ldots \mathbf{r}_N; \dot{\mathbf{r}}_1, \dot{\mathbf{r}}_2, \ldots \dot{\mathbf{r}}_N; t)$, and then

$$Q_j = -\frac{\partial V}{\partial q_j} + \frac{d}{dt}\left(\frac{\partial V}{\partial \dot{q}_j} \right) \qquad j = 1 \text{ to } M \qquad \text{(A6.5.4)}$$

and the Lagrange Equations become

$$\frac{d}{dt}\left(\frac{\partial L}{\partial \dot{q}_j} \right) - \frac{\partial L}{\partial q_j} = Q_j \qquad j = 1 \text{ to } M \qquad \text{(A6.5.5)}$$

where L includes only the conservative forces, and Q_j comprises the non-conservative forces such as frictional forces. (Note that these non-conservative forces are still to be expressed in *functional* form, for example, $Q_j = -\partial\mathcal{F}/\partial\dot{q}_j$ where $\mathcal{F}$ is Rayleigh's dissipation function.) In summary, the non-conservative forces introduce a 'right-hand side' to the Lagrange Equations.

APPENDIX A7.1

Hamilton's Transformation, examples

1) Charged particle moving in an electromagnetic field

In Cartesian coordinates, consider a single particle with electric charge, e, non-relativistic velocity, $\mathbf{v}$ (and speed v), moving through a field with scalar potential ϕ, and vector potential $\mathbf{A}$. The Lagrangian is (with units such that $c = 1$):

$$L = T - V = \frac{1}{2}mv^2 - (e\phi - e\mathbf{A}\cdot\mathbf{v})$$
$$= \frac{1}{2}m\sum_{i=1}^{3}\dot{x}_i^2 - \left(e\phi - e\sum_{i=1}^{3}A_i\dot{x}_i\right) \qquad \text{(A7.1.1)}$$

From Hamilton's transformation equation (7.1), (a type of Legendre transformation), we have:

$$p_i = \frac{\partial L}{\partial \dot{q}_i} = \frac{\partial L}{\partial \dot{x}_i} = m\dot{x}_i + eA_i, \quad i = 1 \text{ to } 3 \qquad \text{(A7.1.2)}$$

2) Particle in a central force field, V

Rectangular coordinates:

$$L = \frac{1}{2}m(\dot{x}_1^2 + \dot{x}_2^2 + \dot{x}_3^2) - V, \quad \text{where} \quad V = V(x_1, x_2, x_3), \qquad \text{(A7.1.3)}$$

and we find that

$$p_i = \frac{\partial L}{\partial \dot{x}_i} = m\dot{x}_i \qquad i = 1 \text{ to } 3. \qquad \text{(A7.1.4)}$$

Spherical polar coordinates:

$$L = \frac{1}{2}m(\dot{r}^2 + r^2\sin^2\theta\,\dot{\phi}^2 + r^2\dot{\theta}^2) - V, \quad \text{where} \quad V = V(r) \qquad \text{(A7.1.5)}$$

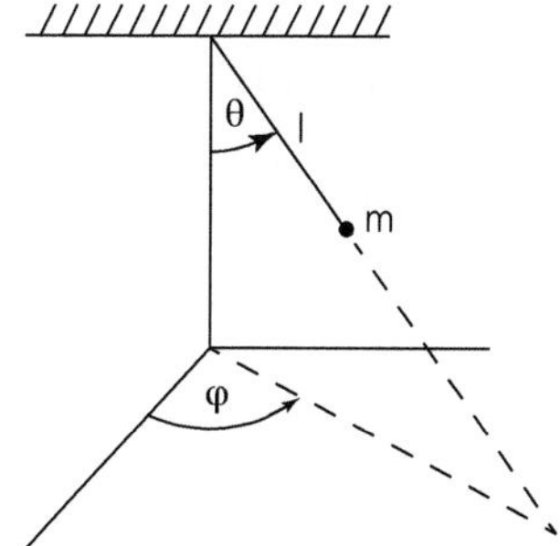

Figure A7.1 Coordinates for the spherical pendulum.

leading to

$$p_r = \frac{\partial L}{\partial \dot{r}} = m\dot{r}, \quad p_\phi = \frac{\partial L}{\partial \dot{\phi}} = mr^2 \sin^2\theta\,\dot{\phi}, \quad p_\theta = \frac{\partial L}{\partial \dot{\theta}} = mr^2\dot{\theta} \tag{A7.1.6}$$

3) Spherical pendulum

$$L = T - V = \frac{1}{2}m[(l\dot{\theta})^2 + (l\dot{\phi}\sin\theta)^2] - mgl(1-\cos\theta) \tag{A7.1.7}$$

leading to

$$p_\theta = ml^2\dot{\theta} \quad \text{and} \quad p_\phi = ml^2\sin^2\theta\,\dot{\phi} = \mathit{constant} \tag{A7.1.8}$$

(as ϕ is 'ignorable').

APPENDIX A7.2

Demonstration that the p_i s are independent coordinates

We can write the Lagrangian as $L = T^{PS} - H$ where H is the Hamiltonian, and T^{PS}, the kinetic energy, has the special form $T^{PS} = \sum p_i \dot{q}_i$. Considering the variation of L, but, for now, the variation of L only with respect to the p-coordinates, then δL is defined as:

$$\delta L = \left[\frac{\partial T^{PS}}{\partial p_1}\delta p_1 + \frac{\partial T^{PS}}{\partial p_2}\delta p_2 + \ldots \frac{\partial T^{PS}}{\partial p_n}\delta p_n \right] - \left[\frac{\partial H}{\partial p_1}\delta p_1 + \ldots \frac{\partial H}{\partial p_2}\delta p_2 + \ldots \frac{\partial H}{\partial p_n}\delta p_n \right] \tag{A7.2.1}$$

However, because of the special form for T^{PS}, each $\partial T^{PS}/\partial p_i$ ends up exactly as $\dot{q}_i$. We can therefore rewrite (A7.2.1) as:

$$\delta L = \sum_{i=1}^{n} \left[\dot{q}_i - \frac{\partial H}{\partial p_i} \right] \delta p_i \tag{A7.2.2}$$

Also, we must remember that we are subject to the canonical equations of motion, $\dot{q}_i = \frac{\partial H}{\partial p_i}$, for all i (equations (7.9) in the main text). This means that the square bracket in (A7.2.2) is always zero. But this then means that each term in the sum in (A7.2.2) is always zero - and zero *irrespective* of the δp_i s. But then δL is also zero irrespective of the δp_i s. The grand conclusion: action is minimized irrespective of the δp_i s, and so the variations of the p_i s don't make any difference, so the p_i s may be considered as truly independent coordinates.

Important note: The T^{PS} above is not the usual quadratic kinetic energy in Cartesian coordinates, it is the kinetic energy in phase space (indicated by the subscript PS). However in phase space there are twice as many 'position' coordinates. This explains why the familiar coefficient $\frac{1}{2}$ no longer appears when we are in phase space.

APPENDIX A7.3

Worked examples in Hamiltonian Mechanics

1) Charged particle moving in an electromagnetic field

We repeat and extend the example started in Appendix A7.1. In Cartesian coordinates, consider a single particle with electric charge, e, non-relativistic velocity, $\mathbf{v}$ (and speed v), moving through a field with scalar potential ϕ, and vector potential $\mathbf{A}$. These potentials are functions of position *and time* (they give the field at the instantaneous position of the moving particle). The Lagrangian is (in units where the speed of light, c, equals 1):

$$\begin{aligned} L = T - V &= \frac{1}{2}mv^2 - (e\phi - e\mathbf{A}\cdot\mathbf{v}) \\ &= \frac{1}{2}m(\dot{x}^2 + \dot{y}^2 + \dot{z}^2) - e[\phi - (A_x\dot{x} + A_y\dot{y} + A_z\dot{z})] \end{aligned} \quad \text{(A7.3.1)}$$

From Hamilton's transformation equation (7.1), we have:

$$\begin{aligned} p_x &= \frac{\partial L}{\partial \dot{x}} = m\dot{x} + eA_x \\ p_y &= \frac{\partial L}{\partial \dot{y}} = m\dot{y} + eA_y \\ p_z &= \frac{\partial L}{\partial \dot{z}} = m\dot{z} + eA_z \end{aligned} \quad \text{(A7.3.2)}$$

(Note that the momentum is not just mass × speed, there is an extra term, eA_i.) The Hamiltonian is given by:

$$H = \mathbf{p}\cdot\mathbf{v} - L = \frac{1}{2}mv^2 + e\phi \quad \text{(A7.3.3)}$$

This is a scalar and is equal to a kinetic energy term plus a potential energy term, as expected. Furthermore, H is equal to the total energy of the system (this is not true for every problem in mechanics, although it is always the case that H has units of energy). While the formulation, (A7.3.3), is correct, it will not lead on to the canonical equations of motion - for that we need to re-express

H just in terms of p- and q-coordinates (remember from the main text that $H = H(q, p)$). The most useful form for H is therefore:

$$H = \frac{1}{2m}(\mathbf{p} - e\mathbf{A})^2 + e\phi \tag{A7.3.4}$$

From Hamilton's canonical equations, (7.9), we obtain:

$$\begin{aligned}
\dot{x} &= \frac{\partial H}{\partial p_x} = \frac{1}{m}(p_x - eA_x) \\
\dot{y} &= \frac{\partial H}{\partial p_y} = \frac{1}{m}(p_y - eA_y) \\
\dot{z} &= \frac{\partial H}{\partial p_z} = \frac{1}{m}(p_z - eA_z)
\end{aligned} \tag{A7.3.5}$$

and

$$\begin{aligned}
\dot{p}_x &= -e\frac{\partial \phi}{\partial x} + \frac{e}{m}\left[(p_x - eA_x)\frac{\partial A_x}{\partial x} + (p_y - eA_y)\frac{\partial A_y}{\partial x} + (p_z - eA_z)\frac{\partial A_z}{\partial x}\right] \\
\dot{p}_y &= -e\frac{\partial \phi}{\partial y} + \frac{e}{m}\left[(p_x - eA_x)\frac{\partial A_x}{\partial y} + (p_y - eA_y)\frac{\partial A_y}{\partial y} + (p_z - eA_z)\frac{\partial A_z}{\partial y}\right] \\
\dot{p}_z &= -e\frac{\partial \phi}{\partial z} + \frac{e}{m}\left[(p_x - eA_x)\frac{\partial A_x}{\partial z} + (p_y - eA_y)\frac{\partial A_y}{\partial z} + (p_z - eA_z)\frac{\partial A_z}{\partial z}\right]
\end{aligned} \tag{A7.3.6}$$

In vector form, this is

$$\dot{\mathbf{p}} = -e\nabla\phi + e\nabla(\mathbf{v} \cdot \mathbf{A}) \tag{A7.3.7}$$

(Knowing the identity, $\mathbf{v} \times (\nabla \times \mathbf{A}) = \nabla(\mathbf{v} \cdot \mathbf{A}) - (\mathbf{v} \cdot \nabla)\mathbf{A}$, from vector calculus, and knowing that $\mathbf{E} = -\nabla\phi - \frac{\partial \mathbf{A}}{\partial t}$, and $\mathbf{B} = \nabla \times \mathbf{A}$, the usual Lorentz force law can be recovered, $\dot{\mathbf{p}} = \mathbf{F}_{Lorentz} = e(\mathbf{E} + \mathbf{v} \times \mathbf{B})$.)

2) Particle of constant mass, m, in a central force field, V

Cartesian coordinates:

$$L = \frac{1}{2}m(\dot{x}_1^2 + \dot{x}_2^2 + \dot{x}_3^2) - V, \quad \text{where} \quad V = V(x_1, x_2, x_3), \tag{A7.3.8}$$

and we find that

$$p_i = \frac{\partial L}{\partial \dot{x}_i} = m\dot{x}_i \qquad i = 1 \text{ to } 3. \tag{A7.3.9}$$

The Hamiltonian is:

$$H = \frac{1}{2m}(p_1^2 + p_2^2 + p_3^2) + V(x_1, x_2, x_3) \tag{A7.3.10}$$

and the canonical equations, (7.9), are:

$$\dot{x}_1 = \frac{p_1}{m}, \quad \dot{x}_2 = \frac{p_2}{m}, \quad \dot{x}_3 = \frac{p_3}{m} \tag{A7.3.11}$$

and

$$\dot{p}_1 = -\frac{\partial V}{\partial x_1}, \quad \dot{p}_2 = -\frac{\partial V}{\partial x_2}, \quad \dot{p}_3 = -\frac{\partial V}{\partial x_3}, \tag{A7.3.12}$$

(Differentiating equations (A7.3.11) with respect to time we arrive at $\dot{p}_i = m\ddot{x}_i$, and noting that $m\ddot{x}_i = F_i$, we return to $F_i = -\partial V/\partial x_i$, the standard result for a central potential.)

Spherical polar coordinates:

$$L = \frac{1}{2}m(\dot{r}^2 + r^2 \sin^2\theta\, \dot{\phi}^2 + r^2\dot{\theta}^2) - V, \qquad \text{where} \quad V = V(r) \tag{A7.3.13}$$

leading to

$$p_r = \frac{\partial L}{\partial \dot{r}} = m\dot{r}, \quad p_\phi = \frac{\partial L}{\partial \dot{\phi}} = mr^2 \sin^2\theta\, \dot{\phi}, \quad p_\theta = \frac{\partial L}{\partial \dot{\theta}} = mr^2\dot{\theta} \tag{A7.3.14}$$

and therefore also to

$$\dot{r} = \frac{p_r}{m}, \quad \dot{\phi} = \frac{p_\phi}{mr^2 \sin^2\theta}, \quad \dot{\theta} = \frac{p_\theta}{mr^2} \tag{A7.3.15}$$

The Hamiltonian is given by:

$$H = \sum_{i=1}^{3} p_i\dot{q}_i - L = p_r\dot{r} + p_\phi\dot{\phi} + p_\theta\dot{\theta} - L \tag{A7.3.16}$$

therefore, after substituting in from (A7.3.15):

$$H = \frac{p_r^2}{2m} + \frac{p_\phi^2}{2mr^2 \sin^2\theta} + \frac{p_\theta^2}{2mr^2} + V(r) \tag{A7.3.17}$$

(Alternatively, we could have arrived directly at (A7.3.17) with the foreknowledge that $H = T + V$ in time-independent cases, and where T is in 'quadratic form'.) Forming the canonical equations of motion (7.9), we arrive at

$$\dot{r} = \frac{\partial H}{\partial p_r} = \frac{p_r}{m}, \quad \dot{\phi} = \frac{\partial H}{\partial p_\phi} = \frac{p_\phi}{mr^2 \sin^2\theta}, \quad \dot{\theta} = \frac{\partial H}{\partial p_\theta} = \frac{p_\theta}{mr^2} \tag{A7.3.18}$$

and

$$\begin{aligned} \dot{p}_r &= -\frac{\partial H}{\partial r} = \frac{p_\theta^2}{mr^3} + \frac{p_\phi^2}{mr^3 \sin^2\theta} - \frac{\partial V}{\partial r} \\ \dot{p}_\theta &= -\frac{\partial H}{\partial \theta} = \frac{p_\phi^2 \cos\theta}{mr^2 \sin^3\theta} - \frac{\partial V}{\partial \theta} \\ \dot{p}_\phi &= -\frac{\partial H}{\partial \phi} = -\frac{\partial V}{\partial \phi} \end{aligned} \tag{A7.3.19}$$

Finally, knowing that we have a central potential (that is, $V \neq V(\phi)$ and $V \neq V(\theta)$) we have:

$$\begin{aligned} \dot{p}_r &= -\frac{\partial H}{\partial r} = \frac{p_\theta^2}{mr^3} + \frac{p_\phi^2}{mr^3 \sin^2\theta} - \frac{\partial V}{\partial r} \\ \dot{p}_\theta &= -\frac{\partial H}{\partial \theta} = \frac{p_\phi^2 \cos\theta}{mr^2 \sin^3\theta} \\ \dot{p}_\phi &= -\frac{\partial H}{\partial \phi} = 0 \qquad \text{and therefore} \quad p_\phi = \textit{constant} \end{aligned} \tag{A7.3.20}$$

We could have predicted that p_ϕ (the angular momentum for rotations within a plane of fixed θ) would be constant, as ϕ does not appear in the Hamiltonian - it is 'absent', indicating that the system is symmetrical with respect to changes in ϕ (but $\dot{\phi}$ does still appear).

APPENDIX A7.4

Incompressibility of the phase fluid

Consider a real fluid in three dimensions, x, y, z. In the 'field description' of the fluid the speeds $\dot{x}, \dot{y}, \dot{z}$, are functions u, v, w, respectively:

$$\begin{aligned} \dot{x} &= u(x, y, z, t) \\ \dot{y} &= v(x, y, z, t) \\ \dot{z} &= w(x, y, z, t) \end{aligned} \tag{A7.4.1}$$

These speeds are components of the velocity vector, $\mathbf{v}$. The condition of incompressibility is:

$$\text{div}\,\mathbf{v} = \frac{\partial u}{\partial x} + \frac{\partial v}{\partial y} + \frac{\partial w}{\partial z} = 0 \tag{A7.4.2}$$

For the phase fluid in $2n$-dimensional phase space (as oppose to a real fluid in everyday space), the generalization of (A7.4.2) is:

$$\text{div}\,\mathbf{v} = \sum_{i=1}^{n} \left(\frac{\partial \dot{q}_i}{\partial q_i} + \frac{\partial \dot{p}_i}{\partial p_i} \right) = 0 \tag{A7.4.3}$$

Now, from the canonical equations (7.9), we know that $\dot{q}_i = \partial H/\partial p$ and $\dot{p}_i = -\partial H/\partial q$. Substituting these into (A7.4.3) we obtain:

$$\text{div}\,\mathbf{v} = \sum_{i=1}^{n} \left(\frac{\partial^2 H}{\partial q_i \partial p_i} - \frac{\partial^2 H}{\partial p_i \partial q_i} \right) = 0 \tag{A7.4.4}$$

However, as we are dealing with 'regular' functions (functions that are continuous, finite, and twice-differentiable), the order of differentiation doesn't make a difference, and we can be assured that $\frac{\partial^2 H}{\partial q_i \partial p_i} = \frac{\partial^2 H}{\partial p_i \partial q_i}$. But this then means that the summation and hence the divergence is automatically zero. In other words, the phase fluid is automatically incompressible, whatever are the functions H, q_i, and p_i (provided they are 'regular').

APPENDIX A7.5

Energy conservation in extended phase space

We consider a system in $2n$ dimensions where the Hamiltonian does depend explicitly on time, t. However, we straightaway re-brand t as the $(n+1)$th position coordinate, $t \mapsto q_{n+1}$. Then, having lost our independent variable, we introduce a replacement independent parameter, τ. We already have the notational convention $dq_i/dt = \dot{q}_i$, and we adopt as well the notational convention $dq_i/d\tau = q_i'$. This means that the speed variables, $\dot{q}_i$, become in the new parameterized version, q_i'/q_{n+1}'. (This follows from: $\dot{q}_i = dq_i/dt = dq_i/d\tau \times d\tau/dt = q_i' \times d\tau/dt = q_i'/(dt/d\tau) = q_i'/q_{n+1}'$.) Note that for the $(n+1)$th coordinate, this means that $\dot{q}_{n+1}$ becomes $q'_{n+1}/q'_{n+1} = 1$. Finally, dt is transformed to $(dt/d\tau) \times d\tau = t'd\tau = q_{n+1}'d\tau$. Making all the substitutions we find that the new parameterized action integral is:

$$A = \int_{\tau_a}^{\tau_b} L^{para}\, d\tau = \int_{\tau_a}^{\tau_b} [L\, q_{n+1}']\, d\tau \qquad \text{(A7.5.1)}$$

where

$$L = L\left(q_1, q_2, \ldots q_{n+1}; \frac{q_1'}{q_{n+1}'}, \frac{q_2'}{q_{n+1}'}, \ldots \frac{q_n'}{q_{n+1}'}\right) \qquad \text{(A7.5.2)}$$

There are two comments to make: L^{para} does not depend on τ - so the system is conservative; no dependence on τ is equivalent to no dependence on q_{n+1}, in other words, q_{n+1} is an 'absent' or 'ignorable' coordinate - but this then means that the conjugate momentum associated with q_{n+1} (that is, p_{n+1}) is a constant (see Sections 6.9 and 7.6 in the main text.).

This makes us ask - what *is* the form of L^{para} in phase space? (Equation (A7.5.2) is in configuration space.) Scrapping the 'speed variables' and substituting instead the p coordinates (see Section 7.3, main text), we will then be in a space of $2n+2$ variables (not only $n+1$ q-variables but also $n+1$ p-variables). We call this the "extended phase space". A curious thing happens in this extended phase space: it is found that L^{para} becomes identical with its kinetic term, and the Hamiltonian term vanishes:

$$\begin{aligned} L^{para} &= \sum_{i=1}^{n+1} p_i q_i' - H \\ &= \sum_{i=1}^{n+1} p_i q_i' \quad \text{and} \quad H = 0 \end{aligned} \tag{A7.5.3}$$

(This happens because L^{para} is a first-order homogeneous function - see Lanczos, page 187.) It seems nonsensical to have a zero Hamiltonian, but there are two ways around this impasse.

(i) Apart from stating that it is a constant, we still haven't defined the conjugate momentum, p_{n+1}. If we choose to equate it to minus the Hamiltonian, $p_{n+1} = -H$, we recover the usual action integral *with* a Hamiltonian (we must also revert back to unparameterized space):

$$A = \int_{t_a}^{t_b} \left[\sum_{i=1}^{n} p_i \dot{q}_i + p_{n+1}\dot{q}_{n+1} \right] dt = \int_{t_a}^{t_b} \left[\sum_{i=1}^{n} p_i \dot{q}_i - H \right] dt \tag{A7.5.4}$$

(Take note of the change in the limits of the sum between (A7.5.3) and (A7.5.4). Also, remember how earlier on we found that $\dot{q}_{n+1} = 1$.)

(ii) Alternatively, we stay in parameterized 'extended phase space', and introduce a condition, K, connecting all the coordinates:

$$K = K(q_1, q_2, \ldots, q_{n+1}; p_1, p_2, \ldots, p_{n+1}) = 0 \tag{A7.5.5}$$

This extra condition can then be introduced into the action integral, using the method of Lagrange Multipliers (Appendix A6.4):

$$\bar{A} = \int_{\tau_a}^{\tau_b} \left[\sum_{i=1}^{n+1} p_i q_i' - \lambda K \right] d\tau \tag{A7.5.6}$$

Without loss of generality, we can set the 'Lagrange multiplier' λ, equal to 1 (as we don't know the value of K, the value of λ can be absorbed into K). We then obtain:

$$\bar{A} = \int_{\tau_a}^{\tau_b} \left[\sum_{i=1}^{n+1} p_i q_i' - K \right] d\tau \tag{A7.5.7}$$

This is the familiar action integral but now in extended phase space (there are p and q variables, $2n + 2$ of them altogether), and K does not contain τ - the system is *conservative*.

We can even go further and merge these two 'work arounds'. From (i) we have $p_{n+1} = -H$, that is, $p_{n+1} + H = 0$; from (ii) we have $K = 0$. We combine these by setting: $K = p_{n+1} + H = 0$. Now K is in effect our new Hamiltonian (in extended phase space) - we call it the 'extended Hamiltonian'.[1] We can insert this K into the action integral (A7.5.7), and minimize it in the usual way, in other words, solve $\delta\bar{A} = 0$. We end up with the usual canonical equations of motion (but now in extended phase space):

Parametric formulation of the canonical equations

$$q_i' = \frac{\partial K}{\partial p_i} \quad \text{and} \quad p_i' = -\frac{\partial K}{\partial q_i} \qquad i = 1 \text{ to } n+1 \tag{A7.5.8}$$

This is the most advanced form for the canonical equations. Some interesting results emerge.

- Our normalization of λ to 1 means that $q_{n+1}' = 1$ which in turn means that $q_{n+1} = \tau = t$. We have p_{n+1} equal to minus the total energy, H, but p_{n+1} is also the 'momentum' conjugate to q_{n+1}. Putting these facts altogether, we find that *the time and minus the total energy are conjugates of each other.*
- After 'disguising' time as a position coordinate, the 'extended Hamiltonian' doesn't depend on the independent variable, τ, and so *every system becomes conservative*. The phase fluid in extended phase space is therefore *static*,[2] and every 'particle' of the fluid remains permanently on a definite hyper-surface, $K =$ constant. (As the value '0' is certainly an example of a constant, then condition (A7.5.5) is satisfied.)
- Finally, it can be shown that all the traditional Principles of Least Action (that due to Euler and Lagrange for t-independent systems, that due to Jacobi, and the one known as Hamilton's Principle) are equivalent to each other,[3] and differ only because of different choices for the condition, $K = 0$.

[1] Goldstein dubs it 'the Kamiltonian' - see Goldstein H, *Classical Mechanics*, second edition, Addison-Wesley (1980).

[2] (there is still motion in configuration space, that is, the speeds are not all zero, but at every position q, there is a *fixed* p).

[3] Lanczos, page 191.

APPENDIX A7.6

Link between the action, S, and the 'circulation'

1) The action function, S, is a function of coordinates q_i and Q_i, for $i = 1$ to n. Now S has the special property that the change in action, ΔS, between a given starting state (q_i^{start}, Q_i^{start}), and a given final state, (q_i^{final}, Q_i^{final}), depends only on these end-states and not on the route connecting them:[1]

$$\int_{start}^{final} dS = \Delta S \quad \text{irrespective of route} \tag{A7.6.1}$$

It then follows that if the start- and final-states coincide, then ΔS will be zero:

$$\oint dS = 0 \tag{A7.6.2}$$

This equation brings to mind the 'circulation' of the phase fluid (equation (7.11)), as it also concerns a closed line integral:

$$\oint \sum_{i=1}^{n} p_i dq_i = constant \tag{A7.6.3}$$

The 'circulation' is not only a constant, it is invariant - that is to say, it is the *same* constant irrespective of the choice of coordinates. This constant is conventionally symbolized, Γ. Thus, invariance of the circulation (for a given system) implies:

$$\oint \sum_{i=1}^{n} p_i dq_i = \oint \sum_{i=1}^{n} P_i dQ_i = \Gamma \tag{A7.6.4}$$

Therefore the difference between the two circulations is zero:

$$\oint \left(\sum_{i=1}^{n} p_i dq_i - \sum_{i=1}^{n} P_i dQ_i \right) = 0 \tag{A7.6.5}$$

[1] This can only be guaranteed if the states are 'close', that is, if ΔS is 'small'.

From equations (A7.6.2) and (A7.6.5), we obtain

$$\oint\left(\sum_{i=1}^{n} p_i dq_i - \sum_{i=1}^{n} P_i dQ_i\right) = \oint dS \tag{A7.6.6}$$

and therefore

$$\sum_{i=1}^{n}(p_i dq_i - P_i dQ_i) = dS \tag{A7.6.7}$$

Finally, by passing from 'differentials' to 'variations', we arrive at:

$$\sum_{i=1}^{n}(p_i \delta q_i - P_i \delta Q_i) = \delta S \tag{A7.6.8}$$

(We are entitled to go from 'differentials' to 'variations' only because S is a 'regular'[2] function, and because the variations are 'small'.)

2 It is finite, continuous, and differentiable.

APPENDIX A7.7

Transformation equations linking p and q via S

From Appendix A7.6 we know that

$$\sum_{i=1}^{n}(p_i\delta q_i - P_i\delta Q_i) = \delta S \tag{A7.7.1}$$

but also δS is a perfect differential and so it satisfies:

$$\delta S = \sum_{i=1}^{n}\left(\frac{\partial S}{\partial q_i}\delta q_i + \frac{\partial S}{\partial Q_i}\delta Q_i\right). \tag{A7.7.2}$$

(in cases where S is a function of q_i and Q_i). As the variations δq_i and δQ_i are chosen independently, the only way that δS from (A7.7.1) and from (A7.7.2) can be equal is if the coefficients of δq_i are equal between (A7.7.1) and (A7.7.2), and likewise the coefficients of δQ_i are equal between (A7.7.1) and (A7.7.2). That is, we require the coefficients of δq_1 to be equal, the coefficients of δq_2 to be equal, ..., the coefficients of δq_n to be equal, and also the coefficients of δQ_1 to be equal, the coefficients of δQ_2 to be equal, ..., and the coefficients of δQ_n to be equal. So we arrive at

$$p_i = \frac{\partial S}{\partial q_i} \quad \text{for all } i \text{ from 1 to } n \tag{A7.7.3}$$

and

$$P_i = -\frac{\partial S}{\partial Q_i} \quad \text{for all } i \text{ from 1 to } n \tag{A7.7.4}$$

APPENDIX A7.8

Infinitesimal canonical transformations

From invariance of the 'circulation' of the phase fluid we know that

$$\sum_{i=1}^{n}(p_i\delta q_i - P_i\delta Q_i) = \delta S \qquad \text{(A7.8.1)}$$

Also, we know that S is the action function, $A = \int_{t_a}^{t_b} dS$. Finally, as well as satisfying (A7.8.1), and being the action function, S has yet another role - it implicitly defines a canonical transformation, a function that transforms from one set of coordinates to another while satisfying the canonical equations of motion, (7.9), in both the new and old coordinates. We shall now introduce one more feature: we let S depend explicitly on t, and see how this affects the canonical transformation functions.

For example, we consider two action functions, S, and, S', occurring at nearby times, t, and, $t + \Delta t$, thereby implicitly defining the canonical transformations, CT, and, CT$'$, respectively:

S, at time t implicitly defines the transformation:

$$\text{CT}, (P_1, \ldots, P_n; Q_1, \ldots, Q_n) \mapsto (p_1, \ldots, p_n; q_1, \ldots, q_n)$$

while satisfying

$$\delta S = \sum_{i=1}^{n}\left[(P_i\delta Q_i) - (p_i\delta q_i)\right] \qquad \text{(A7.8.2)}$$

S', at time $t + \Delta t$ implicitly defines the transformation:

$$CT', (P_1, \ldots, P_n; Q_1, \ldots, Q_n) \mapsto (p_1 + \Delta p_1, \ldots, p_n + \Delta p_n; q_1 + \Delta q_1, \ldots, q_n + \Delta q_n)$$

while satisfying

$$\delta S' = \sum_{i=1}^{n}\left[(P_i\delta Q_i) - (p_i + \Delta p_i)\delta(q_i + \Delta q_i)\right] \qquad \text{(A7.8.3)}$$

Note that Δt is a very small time-interval, and S', which occurs at $t + \Delta t$, is only a very slightly different function to S, which occurs at t. Therefore: p_1

is very close to p'_1, p_2 is very close to $p'_2, \ldots q_{n-1}$ is very close to q'_{n-1}, q_n is very close to q'_n. These proximities are guaranteed only because we insist that S and S' are *continuous* functions. The implications are that the Δq_i s and Δp_i s are *small* quantities whose product and squares are insignificant, and, crucially, S' edges forwards as t edges forwards.

Now transformations that are canonical form a 'group', known as a Lie group. This means that, as CT' and CT are canonical, then any *composition* of operations will also be in the group and will also be canonical. In particular, the composition $CT' \circ CT^{-1}$ is canonical (CT^{-1} is the inverse of CT and is also in the group - has the property of being canonical). But the composition $CT' \circ CT^{-1}$ happens to perform the following transformation:

$$(p_i, q_i) \mapsto (p_i + \Delta p_i,\ q_i + \Delta q_i), \quad i = 1 \text{ to } n$$

while satisfying

$$\delta(S' - S) = \sum_{i=1}^{n} \left[(p_i + \Delta p_i)\delta(q_i + \Delta q_i) - (p_i \delta q_i) \right] \tag{A7.8.4}$$

Putting this line of enquiry to one side for the moment, we remember that S has yet another identity: it is a 'generating function' (it takes the 'wavefront' of common action from one position to the next to the next, and so on, in configuration space). It has the functional form[1] $S = S(q_i, \ldots, q_n; Q_i, \ldots, Q_n; t)$, and so, to first order, we have

$$\Delta S = \sum_{i=1}^{n} \left[\frac{\partial S}{\partial q_i} \Delta q_i + \frac{\partial S}{\partial Q_i} \Delta Q_i \right] + \frac{\partial S}{\partial t} \Delta t \tag{A7.8.5}$$

Furthermore, in our present case $\Delta Q_i = 0$ for all i (as we start from a *fixed* initial point), and therefore we obtain

$$\Delta S = (S' - S) = \sum_{i=1}^{n} \left[\frac{\partial S}{\partial q_i} \Delta q_i \right] + \frac{\partial S}{\partial t} \Delta t \tag{A7.8.6}$$

We now take the variation on both sides of (A7.8.6), and end up with:

$$\delta(S' - S) = \sum_{i=1}^{n} \left[\delta \left(\frac{\partial S}{\partial q_i} \right) \Delta q_i \right] + \delta \left(\frac{\partial S}{\partial t} \right) \Delta t \tag{A7.8.7}$$

[1] Lanczos, page 217, equation (77.4).

(The Δ s do not get varied.) Then, using the relations $\partial S/\partial q_i = p_i$ (see equations (7.14) or Appendix A7.7), we obtain

$$\delta(S' - S) = \sum_{i=1}^{n} \delta p_i \Delta q_i + \delta\left(\frac{\partial S}{\partial t}\right) \Delta t \tag{A7.8.8}$$

Equating the right-hand sides of (7.8.4) and (7.8.8), we find

$$\sum_{i=1}^{n} \left[(p_i + \Delta p_i)\delta(q_i + \Delta q_i) - (p_i \delta q_i)\right] = \sum_{i=1}^{n} \delta p_i \Delta q_i + \delta\left(\frac{\partial S}{\partial t}\right) \Delta t \tag{A7.8.9}$$

Multiplying out the brackets, throwing away variations of Δ s, and ignoring products of two Δ s, we finally arrive at:

$$\sum_{i=1}^{n} (\Delta p_i \delta q_i - \delta p_i \Delta q_i) = \delta\left(\frac{\partial S}{\partial t}\right) \Delta t \tag{A7.8.10}$$

Lanczos calls this a "remarkable relation":[2] all the coordinates are *relative* coordinates (only Δ s and δ s appear); moreover, all the coordinate intervals are '*small*'; and lastly the t-dependence is neatly collected together in *one* place (on the right-hand side). Why is this so - why doesn't t show up on the left? This is explained in the following way. The left-hand side of (A7.8.10) arises from different 'slices' through the phase fluid (such as (A7.8.2) and (A7.8.3)), and we could label each slice with the time that this snapshot was taken. We would then end up with an infinite sequence of 'photos', each with their own t-number and corresponding S. Time thus shows up as a parameter (the photos can be put in order) but it doesn't bring in a *functional* dependence (we cannot say that $S = 3t^2 + 4$, for example). On the other hand, the right-hand side does have a functional dependence on t (through $S = S(q_1, \ldots, q_n; Q_1, \ldots, Q_n; t)$, and a similar form for $\partial S/\partial t$).

We have just seen how the left side of (A7.8.10) depends on p_i s and q_i s, while the right side depends on $(q_1, \ldots, q_n; Q_1, \ldots, Q_n; t)$. However, we want to have both sides in the same space, that is to say, phase space. We therefore determine all the p_i s from $p_i = \partial S/\partial q_i$ (relations (7.14)), and this means that the p_i s are then given as functions of $(q_1, \ldots, q_n; Q_1, \ldots, Q_n; t)$. We then 'invert' these p_i-functions and obtain the Q_i s as functions of $(q_1, \ldots, q_n; p_1, \ldots, p_n; t)$. Finally, we replace the Q_i s in $\partial S/\partial t$ by these Q_i-functions, and so obtain a new function,

[2] Lanczos, page 218.

say, $-X$, in phase-space coordinates:

$$\left(\frac{\partial S}{\partial t}\right) = -X(q_1, \ldots, q_n; p_1, \ldots, p_n; t) \tag{A7.8.11}$$

Taking variations of both sides we obtain

$$-\delta\left(\frac{\partial S}{\partial t}\right) = \delta X(q_1, \ldots, q_n; p_1, \ldots, p_n; t) = \sum_{i=1}^{n}\left[\frac{\partial X}{\partial q_i}\delta q_i + \frac{\partial X}{\partial p_i}\delta p_i\right] + \frac{\partial X}{\partial t}\delta t \tag{A7.8.12}$$

However we don't allow variation of the time (that is, $\delta t = 0$). Therefore, (A7.8.12) becomes

$$-\delta\left(\frac{\partial S}{\partial t}\right) = \sum_{i=1}^{n}\left[\frac{\partial X}{\partial q_i}\delta q_i + \frac{\partial X}{\partial p_i}\delta p_i\right] \tag{A7.8.13}$$

At last, we have $\delta\left(\partial S/\partial t\right)$ in (p, q) coordinates, and we substitute this form into our "remarkable relation" (A7.8.10):

$$\sum_{i=1}^{n}(\Delta p_i\delta q_i - \delta p_i\Delta q_i) = -\sum_{i=1}^{n}\left[\frac{\partial X}{\partial q_i}\delta q_i + \frac{\partial X}{\partial p_i}\delta p_i\right]\Delta t \tag{A7.8.14}$$

The variations, δp_i, are independent, and also the variations, δq_i, are independent - therefore the coefficients of each δp_i must be equal on both sides of (A7.8.14), and also the coefficients of each δq_i must be equal on both sides of (A7.8.14). This leads to the relations:

$$\begin{aligned}\Delta q_i &= \frac{\partial X}{\partial p_i}\Delta t\\ \Delta p_i &= -\frac{\partial X}{\partial q_i}\Delta t \quad i = 1 \text{ to } n\end{aligned} \tag{A7.8.15}$$

Finally, to reach our place in the main text, equations (7.16), we substitute $-\partial S/\partial t$ back for X and arrive at:

$$\begin{aligned}\Delta q_i &= -\frac{\partial\left[\frac{\partial S}{\partial t}\right]}{\partial p_i}\Delta t\\ \Delta p_i &= \frac{\partial\left[\frac{\partial S}{\partial t}\right]}{\partial q_i}\Delta t \quad i = 1 \text{ to } n\end{aligned} \tag{A7.8.16}$$

APPENDIX A7.9

Perpendicularity of wavefronts and rays

The surface of common action, S, is the wavefront. We consider a point X in configuration space just as the wavefront passes through it. We draw the vector, δq^i, starting at X and lying in the wavefront hyper-surface (in other words, it's a vector in the tangent hyper-plane at X). Now this displacement, δq^i, from X cannot change the value of S, as we stay in a surface of constant S. Therefore we must have (in n dimensions)

Surface Displacement, δq^i, satisfying,

$$\sum_{i=1}^{n} \frac{\partial S}{\partial q^i} \delta q^i = 0 \tag{A7.9.1}$$

There is also a displacement, dq_i, along the trajectory or ray. Now dq_i is proportional to the speed, $\dot{q}_i$

Displacement along Ray, dq_i

$$dq_i = \dot{q}_i \, dt \tag{A7.9.2}$$

In many scenarios, with a suitable choice of metric (see below), we have the conjugate momentum given by $p_i = \dot{q}_i$. We also know, from relations (7.14), that $p_i = \partial S / \partial q^i$. Making the substitution $\dot{q}_i$ for $\partial S / \partial q^i$ in (A7.9.1) we obtain

$$\sum_{i=1}^{n} \dot{q}_i \delta q^i = 0 \tag{A7.9.3}$$

But this also implies:

$$\sum_{i=1}^{n} dq_i \delta q^i = 0 \tag{A7.9.4}$$

(A7.9.4) is the scalar product of two vectors and it is zero - *therefore the two vectors are orthogonal*. In other words, displacements, δq^i, within the wavefront, and displacements, dq_i, along the ray, are perpendicular to each other.

Note that we have used tensor notation (upper and lower indices). Also, the metric in this case, ds^2, is defined by the kinetic energy, T, as follows:

$$T = \frac{1}{2}\left(\frac{ds}{dt}\right)^2 = \frac{1}{2}\sum_{i,j=1}^{n} M_{ij}\dot{q}^i\dot{q}^j \tag{A7.9.5}$$

where the M_{ij} are the 'generalized masses', or moments of inertia.

APPENDIX A7.10

Problems solved using the Hamilton-Jacobi Equation

We consider only cases where the Hamiltonian is independent of time.

1) Harmonic oscillator

We examine once again the one-dimensional harmonic oscillator already treated in Appendix A7.3. 'One-dimension' refers to configuration space, and so in phase space we have two coordinates, $q_i = q$ and $p_i = p$. The Hamiltonian for the system is:

$$H = \frac{1}{2}\left(\frac{p^2}{m} + m\omega^2 q^2\right), \qquad \omega = \sqrt{k/m} \tag{A7.10.1}$$

where a mass m, attached to a spring of negligible mass and spring-constant k, oscillates with frequency ω, and at any instant has momentum p, and displacement $x = q$. At the start, the spring is extended and then released (energy is put into the system) but after this no further forces are applied, and we also assume that there is no dissipation. This means that the Hamiltonian does not depend explicitly on time, and so the Hamilton-Jacobi equation reduces to

$$\frac{\partial S}{\partial t} + H\left(q, \frac{\partial S}{\partial q}\right) = 0 \tag{A7.10.2}$$

We seek an S-function that can be separated into a time-independent part, W, and a time-dependent part, ht,

$$S(q, Q, t) = W(q, Q) - ht \tag{A7.10.3}$$

where h is a constant. It then follows that $\partial S/\partial q = \partial W/\partial q$, and $\partial S/\partial t = -h$, and so the Hamilton-Jacobi equation becomes

$$H\left(q, \frac{\partial W}{\partial q}\right) = h \tag{A7.10.4}$$

Our transformation equations, (7.14), tell us that $p = \partial S/\partial q \equiv \partial W/\partial q$, and this along with the form for H given in (A7.10.1), means that (A7.10.4) becomes

$$\frac{1}{2}\left(\frac{1}{m}\left(\frac{\partial W}{\partial q}\right)^2 + m\omega^2 q^2\right) = h \tag{A7.10.5}$$

We find that h is not just any constant, it is the total energy, E. Rearranging (A7.10.5), and substituting E for h, we obtain

$$\frac{\partial W}{\partial q} = \sqrt{m}\sqrt{2E - m\omega^2 q^2} \tag{A7.10.6}$$

and this can be integrated to give:

$$W = \sqrt{m}\int \sqrt{2E - m\omega^2 q^2}\, dq \tag{A7.10.7}$$

(The constant of integration has been ignored but there is no lack of rigour because it will later on be absorbed into another constant, β.)

We pause to explain a few things before continuing. The Hamilton-Jacobi Equation (A7.10.2), can be looked at in a new way. We could view it as a transformation from an original Hamiltonian H in original coordinates (p, q), to a transformed Hamiltonian H' in transformed coordinates (P, Q), where it just so happens that the transformed Hamiltonian is identically zero:

$$\frac{\partial S}{\partial t} = -H(q, p) \mapsto -H'(Q, P) \equiv 0. \tag{A7.10.8}$$

But this zero H' then means that the transformed canonical equations are:

$$\dot{Q} = \frac{\partial H'}{\partial P} = 0 \quad \text{and}$$
$$\dot{P} = -\frac{\partial H'}{\partial Q} = 0 \tag{A7.10.9}$$

and this in turn implies that $Q =$ *constant*, say, α; and $P =$ *constant*, say, β. Our transformation equations, (7.14), also imply that $P = -\partial S/\partial Q$.

Now, we can choose $\alpha = E$, and this leads to $\partial S/\partial Q = \partial S/\partial E$. Also, as we found that $h = E$ (see above) then $\partial S/\partial E = \partial W/\partial E - t$. Finally, from $P = \beta$ and $P = -\partial S/\partial Q$, we have $\beta = -\partial S/\partial Q = -\partial S/\partial E = t - \partial W/\partial E$. We are now ready to return to (A7.10.7). If we differentiate both sides with respect to E (we can differentiate within the integral sign), then we can set the answer equal to $t - \beta$,

$$\frac{\partial W}{\partial E} = \frac{1}{\omega}\int dq \Big/ \sqrt{\left(\frac{2E}{m\omega^2} - q^2\right)} = t - \beta \tag{A7.10.10}$$

This integral can be solved for q, and we find that

$$q = \left(\frac{2E}{m\omega^2}\right)^{\frac{1}{2}} \sin \omega(t - \beta) \tag{A7.10.11}$$

This is the familiar sinusoidal motion of an oscillating spring. Also, the amplitude when squared is proportional to the energy E (this is as expected - the more energy that's put in, the bigger is the amplitude), and the other constant β, gives the phase of the oscillation.

We can also determine the conjugate 'momentum' p, from the usual transformation equation, $p = \partial S/\partial q$, which in this case is the same as $\partial W/\partial q$. Using (A7.10.6) we obtain

$$p = \partial W/\partial q = \sqrt{2mE - m^2\omega^2 q^2} \tag{A7.10.12}$$

and substituting in q from (A7.10.11) we arrive at

$$p = (2mE)^{\frac{1}{2}} \cos \omega(t - \beta) \tag{A7.10.13}$$

(It is easy to verify that $p = m\dot{q}$, as expected.)

2) Planetary motion in 2-D

We consider, more briefly, Kepler's problem of planetary motion[1] with a central 'force-field' $V(r)$, a planet of mass m, and two ('plane polar') coordinates $q_1 = r$, and $q_2 = \theta$. The problem is assumed time-independent (the masses and gravity aren't changing):

$$\begin{aligned} H(r, \theta, p_r, p_\theta) &= H(r, , p_r, p_\theta) \quad (\theta \text{ is 'absent', and there's no } t) \\ &= \frac{1}{2m}(p_r^2 + p_\theta^2/r^2) + V(r) \end{aligned} \tag{A7.10.14}$$

We assume S is separable into time-dependent and time-independent parts, $S = S_t + W(r, \theta)$, and W is further separable as follows: $W = W_r(r) + W_\theta(\theta)$.

[1] The problem has been adapted from Goldstein H, *Classical Mechanics*, 2nd Edition, Addison-Wesley Publishing Co (1980) pp 454-5.

Therefore (the time-independent version of) the Hamilton-Jacobi equation becomes

$$\frac{\partial S_t}{\partial t} + H(r, \partial W_r/\partial r, \partial W_\theta/\partial\theta) = 0 \qquad \text{(A7.10.15)}$$

Now this equation has two terms each depending on totally different things, and yet always summing to zero. The only way this can be satisfied is by having the two terms equal and opposite and constant. The constant is E, the total energy of the system:

$$\partial S_t/\partial t = -E \quad \text{and} \quad H(r, \partial W_r/\partial r, \partial W_\theta/\partial\theta) = E \qquad \text{(A7.10.16)}$$

Furthermore, as θ is 'absent', then $p_\theta = \partial W_\theta/\partial\theta = \textit{constant}$, and setting this constant to α_θ, (A7.10.14) may be written:

$$\frac{1}{2m}((\partial W_r/\partial r)^2 + \alpha_\theta^2/r^2) + V(r) = E \qquad \text{(A7.10.17)}$$

Rearranging and taking the positive square root,

$$\partial W_r/\partial r = \sqrt{2m(E-V) - \alpha_\theta^2/r^2} \qquad \text{(A7.10.18)}$$

Integration of (A7.10.18) with respect to r, and of $\partial W_\theta/\partial\theta$ with respect to θ, yields:

$$W_r = \int \sqrt{2m(E-V) - \alpha_\theta^2/r^2}\, dr \quad \text{and} \quad W_\theta = \alpha_\theta \theta \qquad \text{(A7.10.19)}$$

(ignoring constants of integration). We are not done yet as $W = W_r + W_\theta$ has still more conditions to satisfy: it must comply with our usual transformation equations connecting conjugate coordinates (cf. equations (7.14) in the main text). In this present problem we have:

$$t + \beta_t = \partial W/\partial E = \int \frac{m\, dr}{\sqrt{2m(E-V) - \alpha_\theta^2/r^2}} \qquad \text{(A7.10.20a)}$$

and

$$\beta_\theta = \partial W/\partial\alpha_\theta = -\int \frac{\alpha_\theta\, dr}{r^2\sqrt{2m(E-V) - \alpha_\theta^2/r^2}} + \theta \qquad \text{(A7.10.20b)}$$

obtained by differentiating within the integral. The β_t and β_θ are constants (initial values of t and θ respectively), and α_θ is also a constant - the orbital angular momentum of the planet, $p_\theta = mr^2\dot{\theta}$, often symbolized as l. Equation (A7.10.20b) can be solved to give θ as a function of r - the 'orbit equation'.

General comment

It may seem strange that in both 1) and 2) the method involves differentiating with respect to constants E, or p_θ, but this is in keeping with the approach of Hamilton's Mechanics: the energy and the orbital angular momentum are not just constants they are *parameters,* and much general wisdom is obtained by running the problem again and again with a different energy or a different angular momentum each time.

APPENDIX A7.11

Quasi refractive index in mechanics

The basic differential equation of geometrical optics, discovered by Hamilton, and expressing Huygens's Principle in infinitesimal form, is the Eikonal Equation:

$$\left(\frac{\partial\phi}{\partial x}\right)^2 + \left(\frac{\partial\phi}{\partial y}\right)^2 + \left(\frac{\partial\phi}{\partial z}\right)^2 = \frac{n^2}{c^2} \qquad \text{(A7.11.1)}$$

where ϕ is the 'wavefront', and n is the refractive index of the optical medium that the light is passing through. For light, $n = c/v$, that is, n is inversely proportional to the speed of light, v, in the medium. Also, by the conventions of vector calculus, the left-hand side of (A7.11.1) may be written $|\nabla\phi|^2$. Therefore altogether the above equation may be re-written as $|\nabla\phi|^2 = 1/(speed)^2$.

In the mechanics case, for a time-independent system in Cartesian coordinates, we have:

$$\frac{1}{2m}(p_x^2 + p_y^2 + p_z^2) + V(x, y, z) = E \qquad \text{(A7.11.2)}$$

We can rearrange this to:

$$(p_x^2 + p_y^2 + p_z^2) = 2m(E - V) \qquad \text{(A7.11.3)}$$

Identifying $(\partial\phi/\partial_i)^2$ with $(p_i)^2$, and comparing (A7.11.1) and (A7.11.3) we can correlate an optical problem to a mechanical problem by defining - in the mechanics case - a 'refractive index' for a 'hypothetical optical medium', as follows:

$$\frac{n_{mech}}{c} = const\sqrt{2m(E - V)} \qquad \text{(A7.11.4)}$$

where *const* is some constant of proportionality.

(Note an important difference between light and mechanics. n_{mech} is proportional to $\sqrt{2m(E - V)}$, that is, it is *directly* proportional to the speed, v, of the mass, m; whereas earlier we stated that n for light was *inversely* proportional to the speed, v, of the light.)

APPENDIX A7.12

Einstein's link between Action and the de Broglie waves

The following is a condensed version from Lanczos's book, "*The Einstein Decade (1905–1915)*", Academic Press Inc (1974), pages 113–5.

In the early quantum theory, in Bohr's model of the atom (proposed by Niels Bohr (1885–1962) in 1913), the electron orbits were assumed to be circular and with quantized energy levels, $E = n \times h\nu$, where n had integer values, $n = 1, 2, 3, \ldots$ (h is Planck's constant). Soon afterwards it was realized that the truly decisive quantity that should be quantized was not energy but *action*:

$$\oint p dq = nh \tag{A7.12.1}$$

Now this result was only valid for circular orbits, that is, orbits with just one degree of freedom (the radius of the orbit). The more general case with many degrees of freedom was tackled by the physicists Sommerfeld, and Wilson. They independently of each other came up with a partial solution - the Hamiltonian had to be 'separable' (there had to be one Hamiltonian for each conjugate pair):

$$H = H_1(q_1, p_1) + H_2(q_2, p_2) + \ldots H_n(q_n, p_n) \tag{A7.12.2}$$

with associated quantization conditions:

$$\oint p_1 dq_1 = n_1 h, \quad \oint p_2 dq_2 = n_2 h, \quad \ldots\ldots \quad \oint p_n dq_n = n_n h \tag{A7.12.3}$$

However Einstein could not believe that the particular coordinates, in which the Hamiltonian was accidentally separable, should have a decisive significance for the physical phenomena - the choice of coordinates should not make a fundamental difference. With characteristic ingenuity, he reformulated the 'Sommerfeld-Wilson' quantization conditions in a radically new way. He postulated that it was the *sum* of these conditions that was the quantity of true physical significance:

$$\sum_{i=1}^{n} \oint p_i dq_i = nh \tag{A7.12.4}$$

where n is again an integer (but nothing to do with the n we had before). Now, we know that $p_i = \partial S/\partial q_i$ (from equations (7.14) in the main text), and so we arrive at:

$$\oint \sum_{i=1}^{n} p_i dq_i = \oint \sum_{i=1}^{n} \frac{\partial S}{\partial q_i} dq_i = \oint dS = \Delta S = nh \tag{A7.12.5}$$

(Note that the order of summing and integration has been swapped between equations (A7.12.4) and (A7.12.5).)

Optional commentary

At the beginning of Appendix A7.6 we had a closed integral over the perfect differential, dS, being zero - doesn't this contradict equation (A7.12.5)? Yes, but in Appendix A7.6 that was the classical world whereas now this is the quantum world; and in this quantum world tiny finite values sometimes occur where before we had zero (cf. the order of carrying out observations on conjugate variables).

Einstein's function S, is the same as the action function S, and is proportional to the phase function $\bar{\phi}$, used in Chapter 7.[1] In other words, equation (A7.12.5), discovered by Einstein in 1917, leads to $\Delta\bar{\phi} = 1, 2, 3, \ldots$, which brings us "directly to the doors of de Broglie's matter waves [discovered in 1923]", and "It seems astonishing that this very beautiful idea of Einstein's,... is not mentioned in any of the early histories of quantum theory, although it is such a vital link..."[2]

[1] Notation: Lanczos changes from the phase function, $\bar{\phi}$, in "*The Variational Principles of Dynamics*", to $\Phi/2\pi$, in "*The Einstein Decade (1905-1915)*".

[2] Lanczos L, "The *Einstein Decade*", page 115.

Bibliography and Further Reading

Arnold V I, *Mathematical Methods of Classical Mechanics*, Springer-Verlag, 1978.

Brillouin L, *Tensors in Mechanics and Elasticity*, Academic Press, 1964.

Coopersmith J, *Energy, the Subtle Concept: the discovery of Feynman's blocks, from Leibniz to Einstein*, Oxford University Press, revised edition, 2015; referred to in this book as Coopersmith, *EtSC*.

Ehlers J, Rindler W, and Penrose R, 'Energy conservation as the basis of relativistic mechanics II', *American Journal of Physics* 33(12) 1965 p 995.

Feynman R P, *Lectures on Physics*, with Leighton and Sands, Addison-Wesley, Reading, MA, Fifth printing, 1970

Feynman R P, *QED: the Strange Theory of Light and Matter*, Princeton University Press, 1985.

Feynman R P, and Hibbs A R, *Quantum Mechanics and Path Integrals*, McGraw-Hill Companies, 1965.

Gignoux C and Silvestre-Brac B, *Solved Problems in Lagrangian and Hamiltonian Mechanics*, Springer, 2009.

Goldstein H, *Classical Mechanics*, Second edition, Addison-Wesley Publishing Company, 1980.

Gray C G, Principle of Least Action, *Scholarpedia* 4(12) 2009 p 8291.

Gray C G and Taylor E F, 'When action is not least', *American Journal of Physics*, 75(5) 2007 pp 434–58.

Greenwood D T, *Classical Dynamics*, revised edition, Dover Books, 1977.

Hamilton W R, Theory of Systems of Rays, *Transactions of the Royal Irish Academy*, 15(1828) pp 69–174 (+ supplement in 1830); On a General Method of expressing the Paths of Light and of the Planets by the Coefficients of a Characteristic Function, *Dublin University Review* (1833) pp 795–826; On the Application to Dynamics of a General Mathematical Method previously applied to Optics, *British Association Report 1834*, publ. 1835 pp 513–8; On a General Method in Dynamics; by which the Study of the Motions of all free systems of attracting or repelling Points is reduced to the Search and Differentiation of one central Relation or Characteristic Function, *Philosophical Transactions of the Royal Society*, part II for 1834, pp 247–308.

Hildebrandt S and Tromba A, *The Parsimonious Universe: shape and form in the natural world*, Copernicus, Springer-Verlag, New York, Inc. 1996.

Lagrange J-L, *Mécanique analytique* 1788, translated by Boissonade and Vagliente, Kluwer Academic, 1997.

Lanczos C, *The Variational Principles of Mechanics*, Fourth Edition, Dover Publications, Inc. (1970); '*Space Through the Ages*', Academic Press (1970); '*The Einstein Decade, 1905–1915*', Academic Press (1974).

Landau L D and Lifshitz E M, *Mechanics*, 3rd edition, Course of Theoretical Physics, vol 1, Pergamon Press, 1976.

Lemons D, *Perfect Form: Variational Principles, Methods and Applications in Elementary Physics*, Princeton University Press, 1997.

Mann P, *Lagrangian and Hamiltonian Dynamics for Chemists*, Oxford University Press, in preparation.

Manton N and Mee N, *The Physical World*, Oxford University Press, 2017.

Meriam J L and Kraige L G, *Engineering Mechanics*, vol 1, Statics, 4th edition, John Wiley & Sons, Inc. 1998.

Misner C, Thorne K, Wheeler J, *Gravitation*, W H Freeman and Co. 1973.

Newton I, *The Mathematical Principles of Natural Philosophy*, (1687), translated 1729 by Andrew Motte.

Neuenschwander D E, *Emmy Noether's Wonderful Theorem*, The Johns Hopkins University Press, 2011.

Schutz B, *Geometrical methods of mathematical physics*, Cambridge University Press, 1980.

Sussman G J and Wisdom J, *Structure and Interpretation of Classical Mechanics*, The MIT Press, 2001 and 2nd edition in 2015.

Synge J J and Griffith B A, *Principles of Mechanics*, McGraw-Hill Book Company, Inc. 1949.

Taylor E F, "A call to action" and many other useful publications on the website: www.eftaylor.com

Wells D, *Schaum's Outline of Lagrangian Dynamics*, McGraw-Hill Inc., 1967.

Wheeler J A, *A Journey into Gravity and Spacetime*, Scientific American Library, 1999.

Index